ARE THEY OUT THERE?

ARE THEY OUT THERE?

BEST SELLING AUTHOR

COLONEL GENE P. ABEL

Library of Congress Control Number: 2025901015
ISBN: 978-1-964686-30-1 (paperback)   978-1-964686-22-6 (ebook)

Although this publication is designed to provide accurate information about the subject matter, the publisher assumes no responsibility for any errors, inaccuracies, omissions, or inconsistencies herein.

Photographs and illustrations in this book are reproduced with the permission of their respective copyright holders. A full list of image credits and permissions appears on page 117. Every effort has been made to contact copyright holders. If there are any errors or omissions, please contact the publisher so the author can make corrections in subsequent editions.

Editors: Deborah Froese, John Irvin, Joshua Owens
Cover and Interior Design: Emma Elzinga

Printed in the United States of America

First Edition

3 West Garden Street, Ste. 718
Pensacola, FL 32502
www.indigoriverpublishing.com

Ordering Information:

Quantity sales: Special discounts are available on quantity purchases by corporations, associations, and others. For details, contact the publisher at the address above.

Orders by US trade bookstores and wholesalers: Please contact the publisher at the address above.

With Indigo River Publishing, you can always expect great books, strong voices, and meaningful messages. Most importantly, you'll always find . . . *words worth reading.*

This book is dedicated to the thousands of devoted scientists, the staff at NASA, the DOD, and in Academia who are trying to uncover the truth about the UFO phenomenon.

In addition, this is for all the people who have risked humiliation by coming forward to tell us what they saw and experienced.

Contents

Introduction

hy did I write this book?

After all, much of its contents have been looked over and even folded into pop culture, influencing everything from short stories to television shows. I wanted to sift through the mountains of evidence and present some of the most credible information to help with the question, ARE THEY OUT THERE?

I bring to the writing of this book my experience as a senior executive, including dean of a community college, support division head at a teaching hospital, bank vice president, and chief operating officer of a large public school system. In addition, I served for thirty years as US Army Officer, five of them on active duty, the other twenty-five in the Army Reserve. A nuclear weapons and finance officer, I graduated from the Army War College and was nominated for General Officer.

From 1965 to 1966, I found myself stationed in Hanau, Germany, as part of the 8" Howitzer Battalion where I served as the nuclear weapons officer. While there, I recall reflecting on the infamous event in Roswell, New Mexico, that took place in 1947, and I wondered how many UFO sightings had occurred since. Given my responsibilities included warhead security, I often wondered, with some level of anticipation, if we might experience any UFO sightings in Germany. Unfortunately—or perhaps fortunately, depending upon one's perspective—there were none during my nearly two years stationed there.

I later read about sightings centered over nuclear facilities and wondered how close I might have come to a first-hand UFO encounter.

Since then, the mystery of UFOs has lingered. The skills I learned over my life, plus my passionate interest and extensive research over the past four years, has been employed to consider the possibility that other intelligent life has been and still is visiting planet Earth.

This book contains some of the most credible evidence about UFOs while warning the reader of many false claims and frauds that exist about extraterrestrials. Because, unfortunately, many of these falsehoods have also become part of pop culture in recent decades. I employed the skills from my civilian and military career, especially from the Army War College, to research UFOs and apply critical thinking to the issue. I have also included my military analysis as to what could result from direct alien contact.

A philosophical question accompanies the evidence presented: given the enormity of the universe, is it really possible that humans are the only intelligent life?

Photographs taken by the Hubble and Webb Space telescopes will help you appreciate the size and complexity of the universe. All this is intended to stimulate thought and interest on one of the most dynamic events possible. Might we ever watch a spaceship land on the lawn of the White House and have aliens approach the door of 1600 Pennsylvania Ave?

I hope you enjoy *Are They Out There,* and that it will help answer some of your questions as well as make you want to learn more about the possibility of intelligent life in the Universe.

Gene P. Abel
COL USA Ret.

1

The Back and Forth

Nothing in the universe is unique and alone, and therefore in other regions there must be other earths inhabited by different tribes of men and different breeds of beast.

– Titus Lucretius, *De Rerum Natura* circa 50BCE.[1]

Humanity has spent ages analyzing the world around us, from the depths of the Marianna Trench to the stars dotting the sky at night. More than 1,500 years after Lucretius pondered the existence of other earths, Galileo created the first telescopes to study the celestial bodies around Earth.[2] Our curiosity drove us to record the highly efficient societies of bees, the migratory patterns of birds with their apparent ability to predict the weather, and whales that communicate over extreme distances with their songs.

Why else would we dive into the ends of the Earth to explore the Marianna Trench and find incredible creatures like the red-tipped tubeworms and ghostly fish that survive in hostile environments humanity never thought life could manage in? We have always strived to know the world we live on. Eventually, we turned our attention to the other worlds around us. Our thirst for knowledge is ever insatiable. It appears we have always been asking, "Are they out there?"

Over the years, many claimed to find evidence of alien society on Earth in answer to that very question, and many of those have been proven wrong. Various publications and the internet, especially, have been plagued with misinformation and fraudulent claims, filling the void left by secrets with a veritable sea of nonsense and filler. Unfortunately, most of the time these scams are publicized for financial gain or as a prank on those that seek answers.

For example, in the 2015 compilation of Zecharia Sitchin's work, *The Anunnaki Chronicles*, Sitchin claimed that the Anunnaki in ancient Sumeria were aliens from a twelfth planet called Nibiru.[3] If true, that would give aliens a formative role in the birthplace of civilization, instructing and allowing humans to flourish in some of our earliest years.

The Anunnaki were allegedly powerful beings capable of influencing nearly all aspects of Sumerian life. Unfortunately, sources differ on their number and appearance, making it difficult to definitively say what they looked like or did beyond the various scrolls depicting them, yet there have been no serious conclusions made by scholars that support Sitchin's claim. It is more likely these figures were simply porwtrayals of people or deities used to explain the unknown without the presence of scientific knowledge, as most ancient religions did.

FIGURE 1.1 *Casing Stone from Great Pyramid.*

Daniel Potter, assistant curator at National Museum of Scotland, with a casing stone from the Great Pyramid of Giza. Photo: National Museums Scotland. *Source: Apollo The International Art Magazine,* Garry Shaw January 30, 2019.

Of course, there is also the unending debate on the creation of the Giza pyramids; one which has not been completely answered thanks to the lack of documentation on how they were built in the first place. However, we have been able to determine what they once looked like. Egyptologists, such as Mohamed Megahed of the Czech Institute of Egyptology, have described the pyramids as once having white limestone encasings that were capped by great capstones covered in an electrum-gold composite metal that gave them a gleaming gem-like appearance.[4] All we can say for certain is that the pyramids were likely built with legions of slave labor.

Yet, even with seemingly endless labor, the engineering question of how the massive stones were moved into place and piled in such a monumental fashion remains a mystery. The best theorized solution suggests some 100,000 laborers used sledges, ropes, and levers to haul the massive stones into place.[5]

FIGURE 1.2: *Rendering of original pyramid construction.*

FIGURE 1.3: *Pyramids as they are today.*

Even with evidence from Greek historians, high profile figures and autodidacts such as Elon Musk[6] and Joe Rogan[7] remain skeptical that even 100,000 laborers could move the 2.3 million, nearly three-ton cut stones to build the pyramids. They believe that the laborers were assisted by extraterrestrials with the advanced technology and knowledge to build those wonders of the world. Skeptics go as far as to cite the geographical positions of the pyramids as proof of their extraterrestrial assistance. The pyramids of Giza are oriented along the cardinal directions, north, south, east, and west, with startling accuracy of less than a degree of error.[8] While there is evidence of ancient Egyptians being able to read the stars with such accuracy, these skeptics believe that the massive structures of the pyramids could not have been built with such accuracy without the help of extraterrestrials with advanced instruments to instruct them. Unfortunately, the search for the truth stretches on, often hindered by a lack of permission for historians to properly investigate the pyramids.

We are, by our nature, beings of curiosity. Human beings hunt for knowledge at every turn of the stone and the stars in the hope we can find out how one thing or another works. Failing that, we tend to fill in knowledge gaps with explanations that make sense in the moment.

When ancient Sumerians could not fathom the thought of a world beyond ours, they may have called the celestial bodies in the sky the *Anunnaki*, gods of tremendous power. The Greek may have also considered the sun appearing to circle Earth as a deity on a chariot towing the sun over their blessed lands. Discovering the truths of the Universe through proven methods was the birth of science as we know it.

Around the world, the most basic levels of science, employ the scientific method:

- Make an observation.

- Question the observation.

- Devise a testable explanation known as a hypothesis.

- Perform the test.

- Overserve test results and repeat the process.

Repetition is key to a sound scientific method. When a scientist's explanation is repeatedly proven by their tests, it is considered a theory—not a theory as in a question of whether something exists, a scientific theory. Scientific theories are hypotheses that have been proven over time and through repeated experiments. Unfortunately, it is common for "theory" to be confused outside of the scientific community.

For example, when we were able to observe space and the stars (thank you, Galileo), it became quickly apparent that the sun was not being hauled around Earth by a shirtless man driving a chariot across the sky as Nordic mythology suggests. The shakeup in world knowledge was enough to drive people to execution and exile under claims of heresy, yet the truth shines today regardless. As the familiar adage goes, the truth hurts.

I bring up the scientific method here as a standard to bear in mind when pointing to "evidence" of aliens having visited Earth in ancient times, such as stories about ancient Sumerians and their Anunnaki gods. There are dozens, if not hundreds, of claims that extraterrestrials visited during ancient times, but none have been substantiated by evidence and experimentation. That is not counting the outright fabrications often found online, now further complicated by increasingly convincing AI-generated images and videos.

One higher profile claim states that DNA testing on the teeth of the Paracas Skulls of Peru were found to be nonhuman.[9] For the uninitiated, the Paracas Skulls are a series of elongated skulls that were presumed to be the result of artificial skull deformation via binding at an early age. The skulls, understandably, appear to be alien at first glance—at least according to a commonly held image of aliens—however, repeated tests of similar skulls indicate human biology.

FIGURE 1.4: *Elongated skull.*

There is one major exception involving a Paracas Skull presented at the privately owned Paracas History Museum in Peru, which is notably *not* connected to an educational institution. Brien Foerster, author of more than a dozen books on Incan and Peruvian ancient history and culture, claims to have sent teeth and hair samples from one of the Paracas Skulls to an anonymous geneticist and received the following results:

It had mtDNA (mitochondrial DNA) with mutations unknown in any human, primate, or animal known so far. But a few fragments I was able to sequence from this sample indicate that if these mutations will hold, we are dealing with a new human-like creature, very distant from Homo sapiens, Neanderthals, and Denisovans.[10]

Providing DNA proof of nonhuman life with elongated skulls found in an ancient society would normally have massive repercussions for human history on Earth, potentially proving that some form of extraterrestrial life visited Earth. Unfortunately, there is a reason Foerster's findings were not shouted from the proverbial rooftops or changed textbooks worldwide: they cannot be confirmed with repeated testing.

In Foerster's claim, there was no mention of a university or peer-reviewed study. No repeated tests were conducted or shown to prove his claim. The confusion is further compounded by claims from April Holloway of Ancient Origins, who alleges the Paracas Skulls have dimensions that make them impossible to be human. According to Holloway, the skulls are supposedly 25 percent larger and 60 percent heavier than the average human skull. Unfortunately, there is no corroborating evidence that supports that claim.

By stark contrast, a 2022 article titled "Raman Spectroscopy and STR Analysis of the Elongated Skulls from the Paracas Mummies of Peru" in the *Journal of Biotechnology & Bioinformatics Research* found that DNA samples from hair and tissue on the Paracas Skulls indicated that they were all virtually identical to comparison human hair samples.[11]

Unlike Foerster's claim, Thornton et al. show their experimental process and thoroughly explain how the samples were collected from Paracas remains unearthed in 2016 and scanned using Raman spectroscopy. All the members of the team are named, and their credentials displayed for peer review and response. These steps in transparency are critical to the scientific process and verification of results by other professionals in their respective fields. When looking into claims made online or on television, checking for this kind of due diligence in the claims is necessary before publishing them as factual. Chances are that those publishing such claims without the necessary process are doing so for entertainment purposes, monetary gain, or an attempt to push some ulterior agenda to the masses.

Another claim made by Foerster about the Paracas Skulls is that they must not be human due to the lack of a sagittal suture, the line in the skull that separates the parietal plates of the skull at birth. According to the Center for Disease Control, the sagittal suture can seal as soon as age seven, but generally closes between the ages of twenty-six and thirty.[12] That suggests the lack of a visible sagittal suture on the Paracas Skull, especially one between those ages, does not prove the skull is extraterrestrial in nature.

There are simply too many claims to list and research to stay on topic for this book. I hope that this knowledge, coupled with arming oneself with media literacy and research, can keep readers prepared for the deluge of claims should they decide to brave the wilds of the internet to research extraterrestrials on their own.

Our curiosity is not limited to the organic remains found in archaeological sites around the world; new questions about humanity's past are brought up seemingly every day. Much like the societies of old, when we cannot immediately figure out how ancient peoples performed amazing feats such as building the pyramids of Giza or the marvelous Stonehenge, people are quick to devise theories explaining the phenomena. Often, these quick-fire explanations lack depth. They are either ignorant of—or outright ignore—real, simple explanations based in science and logic. Theories such as aliens moving huge objects with sonic technology tend to ignore the fact that a device with such power would be more unwieldy than the stones they were trying to move—not to mention the complexities of powering such a machine.

FIGURE 1.5: *Map of Göbekli Tepe.*

The Göbekli Tepe (potbelly hill) in Turkey was thought to be nothing more than a medieval graveyard when it was initially discovered in 1963. It was not until 1994 that German archaeologist Klaus Schmidt discovered that the Göbekli Tepe was actually sixteen megalithic rings of carved stone T-shaped pillars reaching up to sixteen feet high and spread over twenty-two acres.

Schmidt's work eventually revealed that the Tepe may have been humanity's first temple to the gods, as the tools used to fashion the pillars were dated 6,000 years prior to Stonehenge's construction. Many of the pillars bear intricate carvings of animals such as foxes, scorpions, lions, and vultures, while many others are still blank. Further excavation and study revealed people who worshipped the Tepe must have been hunter-gatherers, largely incapable of the massive effort and concentrated work required for such a temple.

FIGURE 1.6: *Megalith ring.*

Of course, in an effort to explain the Göbekli Tepe's construction, archaeologists, such as Graham Hancock, jumped to the conclusion that the people of Göbekli Tepe must have been aided by technologically advanced aliens using lesser technology than those that helped the Egyptians construct their pyramids.[13] However, Hancock's beliefs were inadvertently contradicted by Schmidt himself during the excavation. Where Hancock believed

FIGURE 1.7: *Carving on T-shaped stone.*

hunter-gatherers could not have built such a monument, Schmidt believed that the people of Göbekli Tepe settled and became farmers specifically to create the Göbekli Tepe temple. According to Schmidt, for many years it was believed that such structures were only built after people learned to form settlements but the Göbekli Tepe could actually prove the opposite for examples such as Göbekli Tepe.

Similar to the mysteries surrounding the Göbekli Tepe are those surrounding Inga Stone in Paraiba, Brazil. The Inga Stone is a large rock formation that is 151 feet long and twelve feet high. Mysterious petroglyphs are carved along the side, stumping explorers and scholars for centuries. Dating techniques have placed the carvings at approximately 6,000 years old. Not the Stone itself, but the carvings. According to Telma Costa of the Oxford University History Society, that puts the carvings around the end of the last ice age.

What do the carvings covering the Igna Stone mean, though? An archaeologist named Gabriele Baraldi discovered that the Igna Stone contained Hittite writing that depicted several star constellations and indicated the carvers had a working knowledge of celestial movements. *Ancient Origins* would have readers believe these "celestial images" are messages from extraterrestrials bestowing the mathematical formulas for quantum energy and interstellar travel upon the relatively simplistic people in the area at the end of the last ice age.[14] Costa, however, theorizes that these were observed celestial movements that were documented on the Igna Stone and potentially used for agricultural purposes. Costa wrote that several of the other carvings are similar to plant cycles and descriptions. When coupled with the celestial movements, she wrote:

By constructing their monuments in this way, humans would be able to observe and predict the climatic seasons of the equinoxes and consequently prepare themselves to plant or reserve the foods that would serve their needs during difficult times, although it is necessary to remember that the hunters pastors (who were officially supposed to live at that time) would have neither the time nor the knowledge to do so, not to mention the fact that they are always changing and therefore unable to devote themselves to such specific matters, require of them a place to residence."[15]

Unfortunately, no concrete translations or answers to the mystery of the Igna Stone have been discovered. Researchers continue to attempt to study the Igna Stone, but the area is volatile and insecure. If the Igna Stone is destroyed or irrevocably defaced, we may never learn its true meaning.

Another example of rushing to conclusions without the aid of the scientific method is the mysterious London Hammer discovered in London, Texas, in 1946. More specifically, a rock was found by a couple in 1936 and later broken open by their son, George Hahn. Inside this Ordovician rock, was a metal hammer of relatively modern make (circa mid-1800s) with little rusting on the outside. Ordovician refers to the time the rock was formed in nearly 450 million years ago, clearly making it impossible for a modern forged hammer to have been there. That did not stop Carl Baugh of the Creation Evidence Museum from theorizing that the existence of the hammer in such ancient rock meant that the theory of evolution was debunked and that humans existed in the era of dinosaurs.

FIGURE 1.8: *Image of Inga Stone from a distance.*

FIGURE 1.9: *Closer image of Inga Stone carvings.*

It is important to note that the owner of the London Hammer never allowed the wood attached to the hammer head to be carbon dated or any other experimentation to be conducted.

FIGURE 1.10: *London hammer.*

In this image, the relatively modern forged metal of the hammerhead can be seen set into the rock likely much older than it. The wooden handle still remains; testing of which would allow scientists to verify the age of the hammer. Or at least verify when the handle was fitted to the hammer, which would have been before it was encased in the rock surrounding it.

In the absence of permitted testing, however, there are simpler solutions for the London Hammer's mysterious circumstances. In a 1997 article published in *Paleo*, Gary Kuban wrote, "The concretion itself is not Ordovician. Minerals in solution can harden around an intrusive object dropped in a crack or simply left on the ground if the source rock [in this case, reportedly Ordovician] is chemically soluble."[16] Which, when backed up by experimentation on similar rock, proves his theory is at least possible and therefore plausible. At least more plausible than imagining humans wielding forged hammers in the time of the dinosaurs.

The London Hammer now serves and an interesting mystery and an example of the dangers of jumping to conclusions for the sake of confirmation bias to support a sought after conclusions or personal gain, likely both.

FIGURE 1.11: *Fake astronaut carving.*

Recently added carving of astronaut on cathedral in Salamanca, Spain.

An example of research and insight done correctly is related to the above carving of an astronaut on the New Cathedral of Salamanca, in Spain, a church built in the 1600s. When pictures of the carving spread through the internet like wildfire, all manner of questions sprang up as to how a depiction of an astronaut was carved into such an old Spanish church. Surely that meant a seventeenth-century Spanish sculptor had some miraculous contact or insight into modern day space travel

Simply put? No. The sculpture was added as part of a 1992 renovation project organized by the church.[17] As of 2024, searching for the astronaut sculpture quickly reveals the truth of its origins, but the truth was slower to be revealed and counter the first circulations of the misinformation.

When researchers are allowed the opportunity, the scientific method can be applied to ancient history and archaeology. Matters get complicated when it comes to live recordings of extraterrestrial visits as still or video images. Finding high quality picture or video evidence is hard enough as it is. These alleged visits never seem to happen when convenient for filming. With potentially once-in-a-lifetime occurrences, it is often impossible to achieve the repetition required by the scientific method.

But the interference of government institutions also needs to be considered, as they have been accused of repeatedly collecting and destroying evidence and recordings. The idea that governments seek to obscure recent visitations or recovery of extraterrestrial materials from us was once a conspiracy, then an open secret rarely mentioned in an official capacity, and finally, in 2022, the US Congressional hearing on UAPs brought lots of evidence to light from testimonies made under oath. I would like to take you through my personal highlights of the hearing before walking through a timeline of some of the best documented extraterrestrial encounters. Hopefully, you will see the same patterns of activity that I saw when researching these sightings.

2

Congress, Roswell, and Extra(terrestrials)

In 1947, a report of an alien craft crashing in Roswell, New Mexico, kicked off the modern search for alien life. The US government worked tirelessly to cover up the crash and its aftermath but could do nothing to slow down public speculation at the time. Since that fateful day, tens of thousands of photographs and videos have been published around the world that claim to have spotted proof of aliens visiting Earth. Some appear to be real; others are complete fabrication to fuel the public media machine. There have been reports of military pilots seeing objects moving through air and sea that defy our current understanding of possible physics.

HEARINGS ON UNIDENTIFIED AERIAL PHENOMENA

Better yet, in 2022, and on July 26, 2023, the US Congress held hearings on Unidentified Aerial Phenomena (UAP), which is a term that had recently replaced the Unidentified Flying Object (UFO) moniker, and several US government officials testified under oath that pilots and soldiers had observed UAPs during military exercises.[18] The same hearing detailed several government conspiracies such as the recovery of extraterrestrial genetic material and technology that the US government has heavily utilized and kept secret from the public. The 2023 UAP hearing added some of the most credible evidence that Earth has been visited by aliens that the world has ever seen. The hearing even went as far as to include video and photo evidence of military observations and interactions with these UAPs.

Also on July 26, 2023, the US Congress Oversight and Accountability Subcommittee on National Security, the Border, and Foreign Affairs, chaired by Representative Glenn Grothman of Wisconsin, met to hear testimony on Unidentified Anomalous (or Aerial) Phenomena (UAPs). In the hearing, Representative Robert Garcia stated, "We are dealing with your questions to get to the heart of our faith in government. Faith in our institutions is at an all time low. Partisanship and alternative facts make it too easy to doubt our institutions."[19] The hearing was intended to discuss the lack of government transparency regarding UAPs and potentially discovered extraterrestrial activity around and on Earth as well as bring to light the testimonies of whistleblowers such as David Grusch, a former employee of the US Department of Defense (DOD). According to the subcommittee, the existence of UAPs was considered to "pose no threat to national security and are not worthy of further study."[20] Yet the subcommittee recognized that UAPs were a worldwide issue worthy of everyone's attention and concern.[21] While not proof of truth, it is important to recognize that the witness testimonies given during the hearing were done under oath, or an affirmation to tell the truth. Under US law, lying under oath is a crime known as perjury, which is punishable by imprisonment and hefty fines.

Based on the testimonies of former Navy Lieutenant Ryan Graves and Retired Commander David Fravor, there was no doubt in my mind that both of them believed what they were testifying. Graves led by saying that UAPs were in US airspace at the time of the hearing and events were being "grossly under reported" given the potential threat the US faced from them and frequency of their sightings.[22] He described how reporting is discouraged due to the US government's tendencies to discredit those who claim to have seen UAPs, and others fear reporting them, despite being trained observers of airspace. As the hearing continued, others described how the US government consistently shut down and classified operations or sightings of UAPs in an effort to keep them hidden from the public eye.

For Graves's part, he testified that in 2014, during a training mission ten miles off the coast of Virginia Beach the following took place:

Two F-18 Super Hornets were split by a UAP. The object, described as a dark gray or black cube inside of a clear sphere, came within fifty feet of the lead aircraft and was estimated to be five to fifteen feet in diameter. The mission commander terminated the flight immediately and returned to base.[23]

After the flight, Graves talked about how his squadron reported the incident, but nothing else was done about it. Eventually the sighting of UAPs became so frequent that they were included in mission safety briefings, yet to his knowledge, no official actions were taken to find out what was happening or where the UAPs were coming from.

Unidentified Aerial Phenomena Capabilities

For the safety factor alone, it is incredibly concerning to me that no actions have been taken to study UAPs willing to cut so close to our craft or to educate our forces on how to respond to these potential threats. While there was no evidence or testimony that we have been fired on by extraterrestrials, it is possible that our mutual lack of understanding of one another's capabilities could one day lead to an accident of misunderstanding. It is clear from the hearing itself and the reports that came out of it, that we have no idea what kind of capabilities these UAPs have. They appear to be able to change direction and accelerate in defiance of our understanding of physics. Photo and video evidence presented in the hearing, coupled with the advanced radar data from military and civilian flights, make it obvious that these UAPs have vastly more advanced aerospace capabilities than we do.

When asked if there was any indication that UAPs were potentially collecting information on our nuclear capabilities and technology, Mr. Graves, Mr. Grusch, and Mr. Fravor all answered that it was likely or at least possible that they were doing just that.[24] Given the technological capabilities of UAPs, we should be comforted by the fact they have shown little to no aggression toward humans.

In 2010, CBS News reported on an 1967 event in which ten intercontinental missiles were disabled by an alleged UAP near Malmstrom Air Force Base in Montana. At the time, those ICMs were among our most advanced technologies. If they are capable of disabling our most potent weapons, what kind of message are they trying to send us, if any?

Interestingly, in the same CBS report, the author mentioned that, in 1997, Pentagon spokesperson Kenneth Bacon stated, "We cannot substantiate the existence of UFOs, and we are not harboring the remains of UFOs. I can't be more clear about that."[25]

Considering the events reported in the UAP hearing, and the admission that the US government has indeed harbored not only UFO remains, but biologics, the CBS report lends some truth to the testimonies given in 2023.

It is my belief that if we were to encounter one of these UAPs and the situation escalated to violence, we would not be able to engage effectively with them. Judging from the encounters in Montana and the testimonies at the hearing, these entities may simply be able to disable our weapons or even redirect them with devastating effect. Our aircraft would certainly not be able to outmaneuver such amazing technology, at least not at our current technological level.

Caution Advised

Fortunately, from the testimonies given, the UAPs seem primarily interested in observation, not direct conflict. Should they ever decide to directly contact humankind, the situation would far surpass any one nation's security. A violent conflict with an incredibly powerful extraterrestrial would undoubtedly change the trajectory of world history as much as a peaceful, cooperative interaction would. Access to the UAP's technology that defies our knowledge of physics would put humanity into a new era of technological advancement not seen since the rapid advancement of the computer and microchip.

Inconsistencies such as Bacon's statement on UFOs and the revelations of the UAP hearing provoked the witnesses' concerns about the overuse of the US government's system of classifying documents and events. The secrecy runs so deep that Grusch expressed fear that he and the other witnesses coming forward were putting themselves and their families in danger by doing so. Of course, the idea of violent repercussions for exposing secrets is no stranger to truth or fiction. The US government has an easily found list of assassinations throughout its history.

One of the many topics covered in the UAP hearing was the need for the US government to loosen the screws on classified UAP events so the world would know the risks and discoveries made about UAPs. Doing so would also relieve much of the fear surrounding coming forward with information on UAPs and really open the public conversation about aliens and these unknown aerial phenomena. Even during the hearing, Grusch insisted that there were witnesses who would be willing to testify behind closed doors as long as protections for them were in place. I suspect that even then, those witnesses would still be concerned.

Despite Mr. Bacon being "so clear" that no UFOs were recovered, Mr. Grusch testified that the US government was aware of and in possession of "nonhuman spacecraft" as well as biological remains. Until the US government decides to declassify their documentation of these events, all we are able to do is speculate and wonder why they did so and what we have learned from their discoveries. Instead, curious minds are left to dance around the truth with little to no concrete proof. The confirmation and display of knowledge would go a long way to legitimizing the discourse around extraterrestrial life and UAPs. As it stands, serious discourse is largely discounted as conspiracy theorizing or outright fiction for the sake of entertainment. Many that commit themselves to discussing extraterrestrials are left floundering in their imaginations and written off by the general public as little more than creative entertainers.

No One's Following the Money

The fact that Mr. Grusch was able to confirm that there were several classified projects within the US government that operate above the purview of Congress is incredibly concerning. Such revelations open up a host of concerns not just to the power these projects yield in maintaining US secrecy, but also to the potential for white collar crime surrounding the misplacement of funds for these projects. Given that the funds, which are voted on and distributed by the US Congress, are no longer accounted for when they reach these seemingly unsupervised projects, there is no way to hold any body of power accountable for them.

When considering those funds, it is no wonder that the Pentagon has failed to account for over $3.8 trillion in assets and $4 trillion in liabilities in 2023 alone. When funds are allowed to run rampant like that, the secrecy surrounding them can quickly turn into liabilities and crimes that will likely never be solved. It would seem that the classification of projects surrounding UAPs and other classified projects using congressionally distributed funds is a loss for the American public all around between the loss of knowledge and the loss of unimaginable amounts of funds.

The US government would not have to release all the details of their imaging systems either. Simply releasing the images, videos, and radar data without leaking the methods of collection would be sufficient until they are comfortable with showing their technology or the projects become declassified. Officially published images of the nonhuman spacecraft and biologics recovered would go a long way toward confirming the theories as well as putting some of the more outlandish theories to rest.

DATA FROM CIVILIAN SOURCES

Along with that, there is no dedicated place for logging and analyzing data from non-government or non-military sources. Official collaboration would not only progress the study of these lifeforms and technology but would also destigmatize coming forward with information gathered outside official channels. Measures would have to be put in place to deal with hoax and joke submissions, but those have become easier and easier to spot with professional eyes over the years. Just like any other submission method, guidelines for quality would need to be implemented to make sure submitted data was able to be analyzed properly. My hope is that the UAP hearing will generate more interest in extraterrestrials and spark curiosity in the minds of those that wonder if they are out there

Mr. Graves testified that he was involved with "the TikTok Incident" in which he posted a seven-minute-long video to X (formerly known as Twitter) along with a series of high-definition photos of the UAP he saw. In Mr. Graves' post, he intricately detailed his flight path, the location in the sky of the UAP he witnessed, and how several other major airline pilots corroborated the event after the fact. After giving pertinent details about his flight, he wrote:

...I called out a visual on traffic that was excessively bright and looked like about 80 miles range... and then disappeared visually. I never saw the traffic on TCAS. Then a few minutes later I saw two objects round in shape, one lighted, and one not flying in a formation, just above the horizon at a range I guessed of 120-200 [Nautical Miles].

Graves went on to describe how these objects blinked as brightly as a star over and over with white light and continued to do so for the remainder of his flight. Fortunately, he captured the majority of the encounter on his phone, which for the time had an advanced camera capable of high detail and long zoom for a better picture.[26]

Mr. Graves's story, coupled with photographs and video on the post, are a perfect example of the incredible capabilities humankind would face or learn from depending on our interactions with the entities flying the UAPs. There are no known air or spacecraft that can perform as Graves's UAPs did, jumping from commercial flight altitude to beyond Earth's atmosphere so quickly and without a noticeable propulsion that likely would have been picked up on the video. Given Graves mentioned multiple pilots in the area reported seeing the UAP, such an amazing feat of engineering would have certainly been reported from other angles. Graves's post also highlights the other major problem surrounding the stigma of reporting UAP sightings: fear of repercussions. In the original post, Graves was able to include the audio of the radio traffic, but edited out company details, locations, and radio chatter to protect the identities of those surrounding his post. Yes, the edits protected the privacy of those that did not wish to be party to the post; however, they may have been more willing to corroborate the report if such reports were officially recognized and the stigma surrounding UAP reports was at least slightly alleviated.

CLASSIFICATION SECRECY

While the testimonies given during the UAP hearing speak of more recent projects than the likes of Mr. Grusch were part of, the US may have collected the physical and biological remains of extraterrestrials well before the times of their stories. Unfortunately, there is the US classification of projects and total secrecy surrounding events, such as the infamous 1947 UFO crash.

Then again, the US government may be acting with a sense of caution. The world saw what happened in 1938 when H.G. Wells performed a dramatized version of *War of the Worlds* on the radio; it sent the public into a panic and changed radio broadcasting forever. The full release of the US information on UAPs and extraterrestrials could result in a massive panic and public unrest the likes of which the world has never seen. Or the same release of information could unite the world into thinking beyond our terrestrial borders and political understandings and further humanity as a united species.

With the testimonies of these witnesses fresh in your minds, allow me to walk you through a timeline of incredible extraterrestrial encounters and sightings. I suspect the reader will be able to see the patterns of behavior from world governments, particularly the US government, when it comes to the handling of these events. Imagine how much more we, the public, might know about the universe if the following events were revealed in their entirety and humanity's scientific findings allowed to flourish.

3

Governmental Secrecy and the Roswell Incident

A 1941 UFO crash in Cape Girardeau, Missouri, may have been the catalyst that set the US government onto its path of secrecy surrounding extraterrestrials and their related events that was most famously noted later in the Roswell Incident.

In the 2020 edition of *MO41: The Bombshell Before Roswell*, Paul Blake Smith covered the Missouri event in wonderful detail. Smith's research revealed the precursor to Roswell that few talk about in popular culture. In late April 1941, fifty-two-year-old Reverend William Guy Huffman Sr. was resting in his home in Cape Girardeau when he received a call from an unknown police officer. The officer invited Reverend Huffman to visit the site of an alleged plane crash outside of town and read the last rites for anyone who was either dead or dying. He agreed to go, and the police officer picked him up and drove him to the crash site.[27]

When Huffman and the police officer arrived at around 9:30 p.m., the fire department and numerous residents were already on site. Parts of the area were still burning, so Huffman and the officer parked their car out of harm's way and walked to the crash site.

Rather than the expected airplane and injured passengers, they found a strange-looking craft along with what seemed to be three small bodies resembling children with large heads. The gray-colored bodies had long arms that did not appear to be injured. However, two of the entities were apparently dead, and the third, while still alive, was having difficulty breathing and completely unable to communicate.

Reverend Huffman approached the surviving entity and performed their last rites.

The crashed craft was reported to be shiny with a "burnished nickel color." The craft had no wings and was so smooth that it was bereft of any markings for rivets or welds, just a smooth metal exterior. The interior was the same color as the outside, with three small seats that looked to be made for children of a similar size to the crash victims. Also, inside was a band of symbols and small gauges. The symbols resembled Egyptian hieroglyphs but were not any language those that inspected it recognized. Much like the bodies found outside the craft, there was no sign of why the craft had crashed. There were no electrical storms in the area, and no visible evidence as to why it had come down.

Several locals, including one Garland D. Fronabarger, a newspaper photographer at *The Southeast Missourian Newspaper*, were busy taking pictures with their Speed Graphic cameras. Two men in suits held up the corpses of the entities between them while Fronabarger took out a small Kodak 4-A camera and snapped one picture of them with the body. Soon after, a contingent of the military arrived and ordered everyone to turn over anything they may have taken from the site. The military officials also confiscated the Speed Graphic cameras from the press but did not find the pocket-sized Kodak 4-A Fronabarger used to take the picture of the men and the corpse. Everyone at the site was told, "This never happened," and were sworn to absolute secrecy.

When Reverend Huffman returned home, his family could see that this normally calm man was distraught, shaken. After he collected himself, Huffman sat down with his family and told them about his bizarre experience at the crash site. He told them that after his recounting, he would never speak of it to them again. They just listened in amazement at what the reverend told them.

Roughly two weeks after the crash, a man, believed to be Fronabarger, appeared at the Huffman home. Smith wrote, "He handed Huffman a 3x4 black-and-white photo of the dead creature held between the two men in suits." He said he wanted him to safeguard a copy of the photo and indicated that he had more copies as well.

The photograph remained with the Huffman family until about 1953, when Reverend Huffman agreed to allow a family friend, Walter Wayne Fisk, to borrow the photo for authentication and show it to his parents. Fisk promised to return the photo, but he never did.

Not long after Fisk took the picture, he reenlisted in the military and transferred to another post. Fisk was thought to be affiliated with US Military Intelligence.

The existence of the picture was confirmed by Reverend Huffman's granddaughter, Charlotte Huffman Mann, who said she saw the photo on at least twenty occasions prior to 1953. In 1990, UFO authors named Stanton Freedman and Ryan Wood tracked the aging Fisk down at his home where he denied any knowledge of the photograph.[28]

To date, neither the Huffman copy nor any other copies of the photograph have surfaced. This is also the first example of the US government sweeping in and covering such an event up. Given it happened in 1941, when patriotism and allegiance to the government was at high point, those involved with the crash site probably proudly took the secret with them to their graves.

Remember, in April 1941, World War II was raging across Europe, adding to the United States' attitude toward secrecy. The US knew that Hitler was working on new weapon technology and had recovered evidence of their advancements. One theory at the time was that the Russians had jumped ahead of us in the arms race because they had also gotten hold of the same German material. Interestingly, the MO41 (Missouri 1941) predates the infamous Roswell crash by six years and set the tone for how the US responded to UFO crashes from that point forward. It was only a year after Roswell, in 1948, that the Russians reported UFO sightings and claimed to have shot down a cigar-shaped spaceship near Kapustin Yar, a secret military testing facility.

It is entirely possible that this event was also the kicking-off point for the fervor of public interest surrounding the Roswell event. The US government was certainly sensitized to the idea of covering it up as quickly as possible when it happened. The government's handling of Roswell was in line with their handling of MO41, which may have laid the groundwork for their protocols.

THE ROSWELL INCIDENT: JULY 3, 1947

Long before *Are They Out There?* was published, Roswell 1947 was well cemented into pop culture as the quintessential extraterrestrial event. I have always been fascinated with Roswell, and it, along with the uptick in modern reports and the acknowledgment of UFO/UAPs by the Department of Defense, contributed to my decision to write this book. The UFO crash and subsequent cover-up by the US government in Roswell captured the attention and imagination of inquisitive minds around the world.

Colonel Philip Corso's book, *The Day After Roswell*, contains a huge amount of information, which I have taken the time to try and verify from other sources.[29] I researched the incidents and people mentioned in his book, including the Majestic 12 (MJ-12), a secret group of high-level US government officials and military leaders who were allegedly tasked with investigating and managing anything to do with extraterrestrials. In every case, what Colonel Corso claims in his book appears to be accurate. It is worth noting that Colonel Corso was never physically present at Roswell at the time of the event. Rather, he interviewed those who were.

For the uninitiated, in 1947, the US Army operated two major sites near Roswell, New Mexico: the White Sands Missile Range and Alamogordo. At White Sands, the V2 rockets captured from the Germans were being tested. Alamogordo performed nuclear testing. Both installations were extremely sensitive in nature and heavily scrutinized by the intelligence network of the US Army Air Force. (The US Air Force as we know it was formed in September 1947 after the National Security Act was passed.)

In the days leading up to July 3, a series of severe lightning storms rolled over White Sands. During the storms, White Sands's radar showed multiple blips that began to increase at an alarming rate. In response, the Army's Counterintelligence Corps (CIC) sent agents to White Sands, fearing the Russians were spying on their sensitive sites with unknown craft that snuck below detectable altitudes through either Canada or Mexico. Either occurrence concerned the Pentagon greatly.[30]

The Day After Roswell also shares fascinating details about the days following the incident, such as the numerous radar sightings revealed objects that were able to travel at speeds unheard of at the time for any known aircraft. The readings apparently jumped from stationary to 2,000 miles per hour in approximately one second. Aside from being impossible for any craft at the time, such acceleration would have killed any human pilot with the sheer g-force of the maneuver.

While the exact dates have been obscured, sometime between July 2 and 3, during the thunderstorm, the 509th Composite Group of the Army Air Force that was stationed at White Sands observed a radar blip darting across their displays. Shortly after, an explosion produced a brilliant white light eight to ten miles away from the White Sands base.

At the same moment the explosion lit up the sky, the radar blip disappeared.

The explosion was bright enough to be observed by the townspeople of Roswell—in particular, the local sheriff department and fire departments. They were situated about as far away from the crash as the military base was, but on the opposite side of the crash site.

The Counterintelligence Corps (CIC) at White Sands immediately mobilized and notified the base commander, Colonel William Blanchard, of the incident. Colonel Blanchard dispatched the CIC to the crash site with a full recovery team that included trucks, a flatbed, and a crane. A full complement of military police was also sent to the site.

Separately, the Roswell sheriff and fire departments sent their own teams toward the explosion from their side, knowing the military would likely meet them in the middle. They were correct; however, the military contingent arrived before the Roswell locals. By the time they arrived, the Army Air Force had already cordoned off the area and ordered the Roswell responders to stay behind the barrier and not approach the crash site.

However, one member of the Roswell Fire Department located debris that seemed to be from the crash site. The unknown firefighter picked up the piece, wrapped a cloth around it, and tucked it away in a pocket without the military noticing.

According to eyewitness accounts, a saucer-like craft was partially embedded in the ground with a large crack on the side. Three humanoid figures lay dead near the crash. A fourth humanoid was seen crawling out from the wreckage, and a fifth was seen running away from the site. One of the military police officers on site shouted at the fleeing figure to stop before firing and killing it with his rifle.

The military recovery team immediately collected the wreckage in its entirety and stored it on the trucks they brought with them. The craft itself was covered with a tarpaulin and taken to White Sands. Civilian responders still at the site were ordered to forget all they had seen and never to speak of the event with anyone.

Already, the military response was drastically similar to the MO41 incident—all the way down to a piece of evidence being secured away by a concerned citizen at the site of the crash. Only this time, the locals were less inclined to keep the secret.

Shortly after the incident, Lydia Sleppy, who worked for KOAT Radio in Brunswick, New Mexico, received a call from the regional manager of the radio station, John McBoil. McBoil reported the Roswell crash to Sleppy. Not long after getting the call from McBoil, Sleppy received a call from the FBI ordering her to cease all communications surrounding the Roswell Incident.[31] Despite the FBI's efforts, the local newspaper, *Roswell Daily Record,* ran the story on the front page.

FIGURE 3.1: *Front page, Roswell Daily Record, July 8, 1947.*

MATILDA O'DONNELL MACELROY'S ACCOUNT

Most accounts of the July 1947 Roswell crash covered in *The Roswell Incident* disagree about how many alien bodies were recovered from the crash site. A Dr. Weisberg claimed to have seen six bodies, while another resident, Barney Barnett, claimed to have seen several bodies with more possibly in the craft itself.[32] However, the account of Corpsman Matilda O'Donnell MacElroy, a member of the Army Medical Corps assigned to the 509th Bomber Group claims one alien was alive after the crash.

Corpsman McElroy went to the crash site with a Counterintelligence officer, Sheridan Cavitt. Upon her arrival at the crash site, she discovered one of the aliens was still alive. Similar to the others, this gray-colored alien was about four feet tall with long, thin arms and an enlarged head featuring big black eyes. The being also had three fingers on each hand.

As the corpsman examined this alien, she became aware that it was trying to communicate with her by using mental images. She immediately reported this to Agent Cavitt. After consulting with a superior officer, she was directed to accompany the surviving alien back to the base. MacElroy began a dialog with the alien who identified herself as Airl. Corpsman MacElroy, who was promoted to Senior Master Sergeant, conducted interviews with the alien Airl from July 7 through August 1947, at the Roswell Air Base.

At first, communication was via images, but Airl was quickly able to master English, although all communication was always telepathically. Airl's mastery of English enabled MacElroy to understand the alien much more clearly.

Airl claimed to be from a race called Domain and served as an officer, pilot, and engineer stationed on a base located in the asteroid belt surrounding the Earth. Although Airl had a strong female presence, she did not have a gender. Rather, she was a robot with no internal organs but a circuit system that was controlled by some other intelligent being.

MacElroy's account is contained in the book, *Alien Interview* by Lawrence R. Spencer, and based upon documents provided by MacElroy, including official transcripts of her interviews with Airl. The believability of these interviews, which were based on questions provided by intelligence officials, is enhanced by *Alien Interview's* extensive documentation. Detailed footnotes comprise 164 of its 255 pages.

A small sampling of the questions she was asked, and her answers, provides profound insight into what MacElroy experienced:

Q: Are you injured?

A: No.

Q: Do you need food or water?

A: No.

Q: Do you have any special environmental needs, such as air, temperature, atmospheric chemical content, air pressure or waste elimination?

A: No. I am not a biological being.

Q: What is the weapons capability of your people?

A: Very destructive.

Q: Why did your spacecraft crash?

A: Struck by electrical discharge.

Q: Why was your spacecraft in this area?

A: Investigation of "Burning Clouds" of radiation.

Q: Have you visited Earth previously?

A: Yes—long time before Humans.

Q: Are there other intelligent life forms besides yourself in the Universe?

A: Everywhere—We are the greatest/highest of all.[33]

Lieutenant Walter Haut's Deathbed Confession

Lieutenant General Nathan Twining was dispatched to investigate the Roswell Incident at the request of Brigadier General George Schulgen, Chief of Air Intelligence Requirements Division at the time.[34] Under Twining, General Roger Ramey was directed to create a cover story to explain the incident. Lieutenant Walter Haut was instructed to fabricate the official record of a weather balloon crash near Roswell. In 2002, Lt. Haut, while on his deathbed, confessed that the US Army Air Force had ordered him to create the cover story of the weather balloon.[35] The full, unsealed report of Haut's deathbed confession can be found in Appendix A, but I'll outline the most striking information here.

After Haut confirmed that he was of sound mind by reciting his personal information, he described a meeting with several of his superiors the day after the incident to discuss the crash debris field. He said:

Samples of wreckage were passed around the table. It was unlike any material I had or have ever seen in my life. Pieces, which resembled metal foil, paper thin yet extremely strong, and pieces with unusual markings along their length were handled from man to man, each voicing their opinion. No one was able to identify the crash debris.

While weather balloons are not something the average person handles on a routine basis, at least one of the many personnel at the table should have been able to identify a part of their own equipment. After passing around the debris samples, they discussed a plan to divert public attention away from the northern site by simply acknowledging the *southern* site.

Haut later admitted that the weather balloon cover story was their only option to regain control of the Roswell situation. "Too many civilians were already involved, and the press was already informed."

Clearly making such a clean cover-up as MO41 impossible. Only so many people can keep a secret for long.

Lieutenant Haut went on to confess that he was instructed by Colonel William Blanchard to dictate a press release to radio stations KGFL and KSWS, as well as the local newspapers, *Daily Record* and *Morning Dispatch*. According to Haut, the acknowledgement involved telling these outlets that they were indeed in possession of "a flying disc."

It is interesting that their story began with a flying disc they were unable to identify and later became a simple weather balloon crash. It doesn't line up. Weather balloons are generally filled with a potentially flammable hydrogen gas, but they do not have the material content or detonation potential to fuel a crash-based explosion with enough force to be visible over a ten-mile radius.[36]

To Blanchard's credit, at least a crashed metallic disc follows the description of what the Roswell responders saw when they arrived.

Haut's account also stated that he was able to see the craft after the wreckage had been moved to Building 84 which was usually intended for the storage of a B-29:

It [the wrecked craft] was approx. 12 to 15 feet in length, not quite as wide, about 6 feet high, and more of an egg shape. Lighting was poor, but its surface did appear metallic. No windows, portholes, wings, tail section, or landing gear were visible.

Also from a distance, I was able to see a couple of bodies under a canvas tarpaulin. Only the heads extended beyond the covering, and I was not able to make out any features. The heads did appear larger than normal and the contour of the canvas over the bodies suggested the size of a 10-year-old child. At a later date in Blanchard's office, he would extend his arm about 4 feet above the floor to indicate the height.[37]

Haut's description of the wreckage and the bodies allegedly removed from it matched the eyewitness accounts from the crash site. He later went on to describe how the remains of a weather balloon and radar kite had been substituted for the wreckage. I believe this heightened secrecy was tied to the US's concern with reports of flying discs and the origin of these craft. Much like the concern with foreign government technological advances in 1941, the US was concerned that the Russian government had somehow surpassed US technology by using the technology recovered from the Germans during WWII. Ironically, there were others in the US and beyond that assumed the discs were some kind of top-secret US aircraft, rumors that eventually fueled the fervor surrounding Area 51.[38]

Haut ended his deathbed confession with, "I am convinced that what I personally observed was some type of craft and its crew from outer space. I have not been paid nor given anything of value to make this statement, and it is the truth to the best of my recollection."

Lt. General Twining's memo to Brigadier General George Schulgen corroborated Haut's deathbed confession decades earlier. As with Haut's deathbed confession, the entirety of Twining's memo can be found in Appendix B of this book for transparency. Twining wrote to Schulgen that the discs were "something real and not visionary or fictitious." In the same memo, he roughly matched Haut's description of the disc-shaped craft and its dimensions, albeit not in so much detail. Twining did indicate that the crash may have been due to some kind of natural occurrence such as a meteorite coming down but did not seem to invest much into the idea.

FIGURE 3.2: *Brig. General George Schulgen.*

Brig. General George Schulgen with whom Lt. Gen. Nathan Twining communicated regarding the Roswell Incident response.

FIGURE 3.3: *Lt. General Nathan Twining.*

Lt. General Nathan Twining acknowledging the discs were "something real and not visionary or fictitious."

I find it interesting that Twining's memo is one of the few pieces of evidence from Roswell that discusses the military concerns of a craft capable of "such as extreme rates of climb, maneuverability (particularly in roll), and motion which must be considered evasive when sighted or contacted by friendly aircraft and radar, lend belief to the possibility that some of the objects are controlled either manually, automatically or remotely." Not only does this line indicate that White Sands, and possibly other stations, observed these discs, but that they even tried to make contact with them and were evaded. Going back to the 2023 UAP hearing, this would also indicate the possibility that extraterrestrials have been attempting to observe us well before the surge of reported sightings we have seen in recent years. Twining's memo also reports these craft moving in formations "from three to nine objects," with no audible or visible propulsion systems, just like the UAPs from the hearing and other reports not mentioned during the hearing. Unfortunately, the hearing was only given enough time to cover the testimonies of the three men being questioned; otherwise, they may have been able to cover a much wider breadth of reports.

The Roswell Incident was eventually brought to President Harry Truman's attention. President Truman then authorized the Secretary of Defense, James Forrestal, to establish the top-secret group that came to be known as the Majestic 12, or "MJ-12" for the purpose of investigating and possibly recovering extraterrestrial craft. According to records, Truman instructed Forrestal to route all external communications with the office of the president only.[39]

It should be noted that the official stance of the FBI and US government is that the account of the MJ-12 Project is "bogus" as noted by the FBI notes across the presented pages of the document. According to the FBI, the document was nothing more than an elaborate hoax. According to author Barna William Donovan's 2011 book *Conspiracy Films: A Tour of Dark Places in the American Conscious,* skeptics and UFO researchers alike have deduced the MJ-12 Project to be a hoax at best and an elaborate disinformation campaign fabricated by the US government at worst.[40] Donovan points out

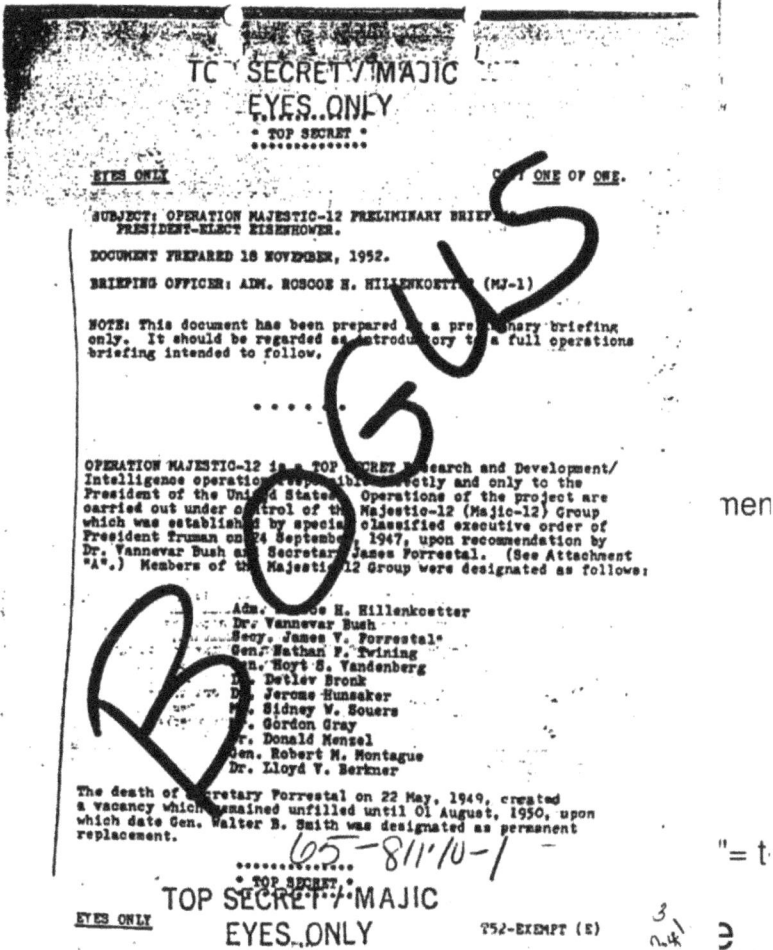

FIGURE 3.4: *Image of page 4 of 24 of FBI MJ-12 document. This is the the MJ-12 document officially released by the FBI detailing the names of those involved with the MJ-12 Project.*

that the document contains several formatting inaccuracies such as the "Top Secret Restricted Information" marking that was not implemented until the Nixon administration in 1969, a full twenty years after Roswell. According to historical documentation, Cutler was not even on the same continent when his meetings with President Truman were supposed to have taken place. UFO researcher and author Timothy Good went so far as to write that the MJ-12 Project was likely a government hoax intended to discredit those that use the documents as proof of the Roswell Incident.

Regardless of claims that the MJ-12 was a hoax, the people listed in the memorandum were very real and did study in the important fields represented in the document. So, who were these majestic twelve? Readers will see some familiar names.

- Vice Adm. Roscoe H. Hillenkoetter: the first director of the Central Intelligence Agency.

- Vannevar Bush: the engineer who marshaled American technology for World War II and ushered in the Atomic Age.

- James V. Forrestal: secretary of the navy and the first US secretary of defense.

- Gen. Nathan Twining: chief of staff of the US Air Force from 1953 until 1957 and chairman of the Joint Chiefs of Staff from 1957 to 1960.

- Gen. Hoyt S. Vandenberg: the second chief of staff of the US Air Force.

- Detlev Wulf Bronk: a prominent American scientist, educator, and administrator credited with establishing biophysics as a recognized discipline.

- Dr. Jerome C. Hunsaker: an aviation pioneer who founded the first college course in aeronautical engineering at the Massachusetts Institute of Technology.

- Sidney William Souers: an American admiral and intelligence expert who was appointed as the first director of central intelligence on January 23, 1946, by President Harry S. Truman.

- Gordon Gray: an official in the government of the US during the administrations of Harry Truman and Dwight Eisenhower who was associated with defense and national security.

- Donald Howard Menzel: one of the first theoretical astronomers and astrophysicists in the US.

- Robert Miller Montague: a lieutenant general in the US Army who achieved prominence as the deputy commander of Fort Bliss, Texas, and commander of the Sandia Missile Base in New Mexico.

Lloyd Viel Berkner: an American physicist and engineer who was one of the inventors of the measuring device that since has become standard at ionosphere stations because it measures the height and electron density of the ionosphere.

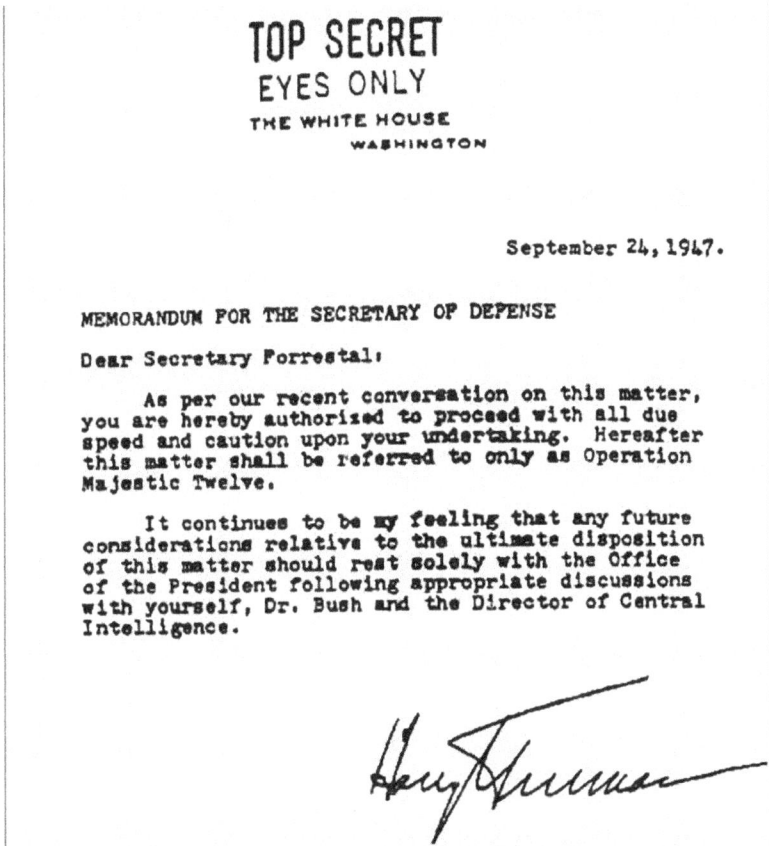

TOP SECRET
EYES ONLY
THE WHITE HOUSE
WASHINGTON

September 24, 1947.

MEMORANDUM FOR THE SECRETARY OF DEFENSE

Dear Secretary Forrestal:

　　　　As per our recent conversation on this matter, you are hereby authorized to proceed with all due speed and caution upon your undertaking. Hereafter this matter shall be referred to only as Operation Majestic Twelve.

　　　　It continues to be my feeling that any future considerations relative to the ultimate disposition of this matter should rest solely with the Office of the President following appropriate discussions with yourself, Dr. Bush and the Director of Central Intelligence.

FIGURE 3.5: *The letter sent to Secretary Forrestal by President Truman.*

I independently verified the extensive claims made by the colonel with respect to all of the figures involved in his book and the 1947 Roswell incident. In each case, the colonel's description of who they were and what they did was verified by several other sources, and it appears that everything he stated about the individuals he identified was factual.

If Forrestal did indeed visit and oversee the recovery and study of the craft and biological remains of extraterrestrials, it would add a layer of concern to reports surrounding his behavior afterward. After Forrestal's visit to Roswell in 1947, there was reportedly a marked change in his demeanor. Just two years later, in 1949, Forrestal resigned from his position. Soon after, his mental and physical condition deteriorated, and he was placed in a naval hospital for approximately one month.

Eventually, Forrestal's brother demanded Forrestal be released from the hospital and informed officials that he was going to be removing him either way. The night after Forrestal's brother issued that demand, James Forrestal fell to his death from a sixteenth story window at the naval hospital.[41]

Given the secrecy surrounding Roswell and his apparent presence for so many important events, is it possible that Forrestal was killed to maintain his secrets? Could Forrestal have had a valuable deathbed confession similar to Lieutenant Haut? Unfortunately, we will likely never know. Or, if we are given some kind of answer, it will be mired in skepticism and doubt surrounding information released by the US government. This is yet another example of how the US' secretive programs have potentially taken lives and obscured important information from the world, potentially stifling significant shifts in public perception of our universe.

Forrestal seemed to believe that the public had a right to know of the US' involvement with UFOs and extraterrestrial discoveries. According to the documents, he was often at odds with other members of the MJ-12. However, there was never any definitive proof that Forrestal's death was a murder or that it was related to his alleged conflicts with the group. Forrestal's death was declared a suicide by naval hospital staff. No further investigation into Forrestal's death took place.

Thirty years after the incident, retired Major Jesse Maxwell, who was at Roswell when the crash occurred, came forward and admitted that the crash did in fact occur and that he had been directed to cover up the incident. Maxwell claims, at the time of the crash, he saw the debris from the ship itself and dead aliens moved into an autopsy facility at White Sands, New Mexico. Maxwell also claims that some of the debris removed from the crash site remained at White Sands along with two of the dead aliens. The ship and three of the other dead aliens were transported to Wright-Patterson Air Force Base.

In 1955, a US Air Force technician by the name of Norma Gardner was stationed at Wright-Patterson Air Force Base. Her job was to catalog items taken from alien crashes. She cataloged more than one thousand items from UFO crashes but was sworn to secrecy about what she saw. Airman Gardner, on her deathbed, said, "The government [could] no longer keep [her] from telling the truth." She explained that not only had she cataloged unidentified items from alien crash sites but that she personally saw four-foot gray beings in tanks at Wright-Patterson Air Force Base. She also claimed there were autopsy reports from the gray beings.[42] Even the popular term for extraterrestrials, "Grays," can be traced back to the Roswell incident.

From 1955, we jump to 1961 and the familiar name of Lieutenant Colonel Philip Corso, who was freshly assigned to head the Foreign Technology Desk in the office of Army Research and Development at the Pentagon. Prior to that assignment, Corso served under General MacArthur in Korea and later on the National Security Council under Dwight Eisenhower, whom Corso knew personally. As the head of the Foreign Technology desk, Lieutenant Colonel Corso reported directly to the Chief of Army Research and Development in the Pentagon, Lieutenant General Arthur G. Trudeau. Corso served in that office from 1961 to 1963.

4

Sightings after Roswell

If the incidents in Missouri and Roswell had been the only occurrences suggesting alien visitors, interest in extraterrestrials would have likely faded away with time. However, many other sightings took place in the US and all around the world, keeping public interest alive and well over the decades. All over the world, people came together to form communities dedicated to independently investigate UFO sightings and evidence of extraterrestrials.

In October 2023, CNN reported that the Pentagon was receiving more than a dozen UAP/UFO sightings per month.[43] Earlier in 2023, the world was abuzz about a series of UAPs that were seen crossing the North American continent, well into the United States of America.[44] Those UAPs turned out to be Chinese spy balloons that bristled with surveillance equipment. Once the balloons were over safer airspace, the US Department of Defense shot them down and collected the remains.

Below is a selection of other UAP sightings over the last seventy years or so that show why interest amongst government and the public has been growing. Because of these increased observations, many scientists, apart from the government, have become interested in investigating UFO sightings. A number of organizations have expended a great deal of time, energy, and money to document and analyze sightings. Some of the most notable follow.

MARCH 1951: PROJECT BLUE BOOK

Authorized by General Nathan Twining and headed by Captain Edward Ruppelt, Project Blue Book was a US Air Force project designed to investigate UFO sightings. The objective of these investigations was to determine whether UFOs were a threat to US national security and to analyze UFO data. The project ran from 1951 until 1969 and collected 12,618 UFO reports.[45] At the end of the project, the Air Force concluded that most of the reports were misidentifications of natural phenomena or conventional aircraft sightings, and that none of the sightings were a threat to national security or worth investigating further. A civilian branch of Project Blue Book continued until 1989 but was terminated with similar conclusions.

However, after looking over the information in Project Blue Book, people, such as author Brad Steiger, concluded that Project Blue Book may have been an effort to discredit and hide real UFO sightings and evidence.[46] While it is possible that many of the sightings reported by Project Blue Book were misidentifications or simply hoaxes, Steiger found evidence that legitimate sightings were buried in the report and not focused on as much as they should have been.

Fortunately, he was able to access the full investigations and reports with the Freedom of Information Act. But even if some of the sightings were civilians spotting advanced military weapon testing, why would the military test such flights in areas that people would be able to see them? Why not keep their testing to restricted areas that observation can be controlled from? The most likely answer is that the "secret testing of advanced weapons or flight systems" was an easy response, one that the general public could never investigate. It provides a dead end instead of the truth.

Several of the 12,618 incidents reported by Project Blue Book were seen by large numbers of people, many of whom were trained observers such as police officer and pilots, who are trained to provide accurate and detailed assessments of any given situation. Photographs and radar readings even backed up some of these sightings. Yet the project claimed to have identified all but 701 observed objects. It is only natural to challenge the credibility of explanations provided by the government. Something like a flying saucer is not likely to be mistaken by even the casual observer for an ignition of swamp gas, flying birds, or a lighthouse beacon sweeping the skyline.

Cover-ups like Project Blue Book point back to the congressional hearing on UAPs mentioned earlier, which drummed up fresh public interest in the idea of extraterrestrial visitors and how we might respond to them if they make contact. With the US government finally acknowledging the existence of unidentified aerial phenomena and their investigations into it, I hope the public will now engage in unbiased discourse about what is happening around our planet. The public needs freedom of access to the last decade of classified documents regarding UAPs. Only then will we know how many encounters have actually occurred. Only then will we obtain insight into who and what is visiting Earth and what, if any, assistance or interference has come from the military.

AUGUST 25, 1951

Three scientists from Texas Tech, in Lubbock, Texas, saw twenty-five to thirty lights across the sky. They saw a second group in the same area and contacted the US Air Force. The objects continued to appear until September 1951.[47]

AUGUST 30, 1951

Carl Hart Jr., a student in Lubbock, Texas, took two photographs of an unidentified flying object. He later provided the pictures to the US Air Force, which ended up in Project Blue Book. The Air Force later claimed they were streetlight reflections over a flock of birds that flew by. But the speed with which these objects moved far exceeded anything a bird could achieve.[48]

December 9, 1965

The Flight Aviation Administration (FAA) tracked the flight paths of several unidentified objects which were verified by twenty-three pilots in the airspace. Not long after, there was a crash in the woods near Kecksburg, Pennsylvania. Several people from Kecksburg found the downed craft, which was described as acorn-shaped with windows on the sides.

John Murphy, a radio host at WHJB in Greensburg, Pennsylvania, reported on the crash. Much like MO41 and Roswell, soon after the crash was reported, the military swept in, told people to leave, and packed up the wreckage on a flatbed truck. The military told bystanders they had witnessed a meteor impact—despite the fact that there were more than one hundred witnesses from the nearby town.

Murphy interviewed all of the witnesses and planned to present what he had learned about the Kecksberg crash on a radio show. However, several men in black suits approached him and convinced him not to air the story. They also seized Murphy's notes and the recordings concerning the incident. In 1967, Murphy reopened his investigation into the crash. Two days after he announced his plans on the radio, he was killed by a hit-and-run driver who was never found.[49]

March 14, 1966

In Washtenaw County, Michigan, more than 100 people near Dexter, Michigan, including Sherrif Deputies Buford Bushroe and John Foster, reported seeing UFOs moving at incredible speeds near town. Bushroe and Foster described an object with flashing lights that changed color frequently between blue, green, and white.[50]

These sightings continued for most of the week until, on March 20, there were reports of a craft hovering around town and possibly landing. Bushroe and Foster reported the event to the US Air Force, who brought in Astronomer J. Allen Hynek to investigate. There were so many reports beyond Bushroe and Foster's that then-Congressman Gerald Ford demanded a hearing to investigate the reported sightings. The US Air Force later claimed people were seeing swamp gas. Hynek later went to work with the Center for UFO Studies in Chicago, Illinois.

JUNE 15, 1968

On a river near Cua Viet, Vietnam, around 12:30 a.m., two US Navy patrol boats, one commanded by Lieutenant Davis and the other by Lieutenant Snyder, saw two flying saucers, dubbed "enemy helicopters" as the enemy did not have helicopters at the time, hovering over the river in front of them. Lt. Davis's patrol boat exploded and was destroyed just before the two UFOs disappeared. Lt. Snyder was ordered to continue moving up the river. Two UFOs appeared again, and Snyder's boat opened fire on them. Immediately, the same kind of fire that he was using against the UFOs was returned at his patrol boat. Snyder decided to turn around and head into open water where the Australian vessel, the *Hobart,* was operating. Snyder reported the incident to the Navy, who dispatched several F-4s from Da Nang. Lt. Snyder was ordered to maintain radio silence.

Once in the area, the Navy F-4s spotted two UFOs in the ocean near the *Hobart.* They fired air-to-air missiles at the two objects, which immediately departed at a high rate of speed. The description of the UFOs near the *Hobart* matched those described by the patrol boats on the river.

The following day, a missile that appeared to come out of nowhere struck the *Hobart* while at sea. Immediately, the crew saw the two UFOs and the *Hobart* was struck by two more missiles, causing extensive damage. When the US Navy examined the damage to the *Hobart,* they found pieces of the missiles that struck the ship contained fragments of serial numbers matching missiles fired at the UFOs the previous day. Allegedly, the UFOs were able to capture and return the missiles in the same manner they returned the rounds from Snyder's patrol boat.

According to the US Navy after-action incident of the report, it was determined that it was unsafe to engage the "enemy helicopters" for fear of similar reactions and damage.[51] It is important to note that there were several technological issues surrounding the *Hobart* incident that could have indicated both the patrol boat and the *Hobart* were subject to incidents of friendly fire. In an already tumultuous warzone under incredible amounts of stress that lead to an unfortunate loss of life for both the US and Australian forces, friendly fire cases were uncommon but not unheard of. According to author Jeffrey Grey in *Up Top: The Royal Australian Navy and Southeast Asian Conflicts,* the *Hobart* was approached and fired on by an unknown craft with no IFF signature on radar. By Grey's account, the missiles were only determined to be of US Air Force make, which would make the incident an unfortunate case of friendly fire. At no point was it mentioned that the *Hobart* was under attack by UFOs.

However, if we accept that the *Hobart* and the US patrol boats were indeed engaging with extraterrestrials, this event stands out at the only one I have read that includes aggression from the UFOs. If true, it is clear that trying to attack these beings with conventional weaponry would be futile at best and deadly for the attackers at worst. Perhaps the invention of Directed Energy Weapons will help if it comes to blows with these visitors, but that is on the assumption that they have not improved their equipment in the time since either. When faced with the increasing devastating capabilities of our weapons, we have to wonder what the outcome would be if the improved weapons were fired back on us without warning as Snyder's patrol boat gunfire was.

January 6, 1969

Former President Jimmy Carter is one of the most prominent people to have witnessed a UFO sighting. The sighting occurred when he was in Leary, Georgia, in 1969, to address the Lions Club. He and the ten to twelve people with him saw the UFO above the horizon. He filed a report with the National Investigations Committee on Aerial Phenomena on September 18, 1973.

Reportedly, he said it was the "dandiest thing" he had ever seen.

When Carter ran for president in 1976, he promised to release any government information on UFOs to the public should he be elected, but he reneged on that promise because of national security concerns.[52]

January 18, 1978

During the early morning, a UFO was spotted flying over the Fort Dix and McGuire Air Force bases in Burlington County, New Jersey. An on-duty military police officer in McGuire Air Force base saw something at the edge of the runway. It turned out to be a small gray-colored being with a large head and a slender body. The MP panicked, took out his .45 sidearm, and shot the being several times. It fled over the fence between the two bases.

The MP contacted Air Force Sergeant Morris and was told to go to the gate. Sergeant Morris and his colleagues found the alien dead along the runway smelling strongly of an ammonia-like odor. Later that day, a team from Wright-Patterson Air Force Base arrived in a C-141 cargo aircraft, put the body in a silver-colored box, and took off.

Morris and his companions were warned by security not to talk about the incident, or they would be court-martialed. Two days later, Morris and the others were taken to Wright-Patterson Air Force Base where they went under extensive interrogation and were again warned not to mention the incident.

Upon return to McGuire, Morris debriefed his commander, Major George Filer. Nothing more was said about the incident until many years later.

In 2001, Filer spoke with ABC News reporters Katelynn Raymer (Washington) and David Ruppe (New York). He told them about the discovery of an alien body that had been shot dead by one of MPs.[53]

November 17, 1986

At 5 p.m. over Alaska, during Japan Airlines flight 1628 from Reykjavík to Anchorage, seasoned pilot, Captain Kenju Terauchi, encountered two unknown craft to his left, south. These craft flew at the same speed as the 747. At 5:18 p.m., the two objects swooped ahead and assumed a stack position in front of the airliner, flying as though they had overcome gravity. Anchorage traffic control could not confirm these aircraft on their short-range radar, nor could other flights in the area.

After three minutes, the objects assumed a side-by-side configuration alongside the plane, which they maintained for another ten minutes, accompanying the airliner. At 5:23 p.m., the two craft abruptly streaked forward and vanished into the eastern horizon. Using his onboard radar, Terauchi confirmed the craft were approximately seven and a half nautical miles distant. He reported the aircraft had moved and changed directions with a speed like nothing he had ever seen.

As the flight neared Fairbanks and the city lights began to illuminate objects in the sky, Captain Terauchi noticed another huge object to his left. The spaceship was approximately twice the size of an aircraft carrier and flying at thirty-one thousand feet.

Terauchi executed a 360-degree turn, hoping the gigantic craft would continue forward. However, the aircraft mirrored his turn and continue to shadow the 747. This continued for thirty-one minutes. The entire sighting of this object lasted approximately fifty minutes.

After Japan flight 1628 landed safely in Anchorage, Captain Terauchi filed a formal FAA report together with the data from his onboard radar and the tapes from the conversation between him and the tower at Anchorage.

Philip J. Klass, editor for the *Aviation Week and Space Technology* magazine, later wrote that Mars and Jupiter were aligned properly to appear where the captain allegedly saw the lights of the craft. Klass also wrote that the lights from Fairbanks filtering through the clouds could explain the larger phenomenon that the captain saw.

Unfortunately, no other craft or the FAA were able to confirm the radar presence of multiple UFOs the captain described, and nearby flight accounts of the encounter differed from the captain's version of events.[54] Klass later wrote a follow-up article for the *Skeptical Inquirer* in which he reviewed the results of the FAA investigation and full data related to said investigation. In Klass's article, he wrote that the FAA was only able to confirm one radar ping, not three UFOs per the pilot's claim. Two other craft in the area, United 69 and a USAF C-130 transport were also coordinated to try and confirm the UFO sighting but were unable to do so. At one point, United 69 was positioned to be directly in front of where Terauchi claimed to see the UFO and reported no contact.[55]

NOVEMBER 29, 1989, TO MARCH 30, 1990

What later became known as the Belgian UFO Wave began on November 29, 1989. It included over eighty locations across Belgium and over 150 similar yet separate reports. All sightings reported brightly lit triangular UAPs with three white lights on each point and a red light in the center. On April 4, 1990, a photograph was allegedly taken of one craft and sent for an extensive year-long examination by Professor Mark Ashwa at the Royal Military Academy in Brussels.[56]

According the 2011 film *Secret Access: UFOs on the Record*, the photograph was determined to be genuine. However, Ben Deighton of *Reuters* also reported that a forger had come forward and claimed, "You can do a lot with a little, we managed to trick everyone with a piece of polystyrene."[57] While the photograph may have been faked, and footage or photographs would have been harder to come by in 1990, how did so many people report such similar observations over such a widespread area?

During the event, lights from these objects were reportedly bright enough to read by. One police officer compared them to the lights over a football field. The sightings lasted for months, far longer than any typical explanation for lights in the atmosphere to come and go, if they were even to be believed.

The Belgian UFO Wave culminated on March 30, 1990. Two Belgian Air Force F-16s pursued the mysterious craft. During the chase, the craft appeared to defy the laws of physics, accelerating from 150 knots (172 mph) to more than 990 knots (1,139 mph) in less than a second.

FIGURE 4.1: *Triangular spaceship.*

Apparently, the pilots were unable to make visual contact with the craft, despite their targets being confirmed by radar and on-board instruments.[58]

FEBRUARY 2001 TO MARCH 2002

While not a UFO sighting, a computer system administrator named Gary McKinnon found potential evidence of such sightings. From February 2001 to March 2002, McKinnon hacked into several networks within the US military and NASA. In these networks, he was able to access Excel sheets containing names and ranks of "non-terrestrial officers" as well high-resolution images of what he described as a "silvery, cigar-shaped object with geodesic spheres on either side. There were no visible seams or riveting . . . The object didn't look man-made or anything like what we have created."[59]

McKinnon's claims are not without reason for skepticism. By his admission, he viewed the high-resolution images after downscaling them enough to view remotely through a 56k connection, which would severely degrade the image quality and context. It is also possible that McKinnon was looking over potential projects that NASA was conceptualizing. Without the context of images documentation, we will never know.

McKinnon's actions certainly rattled the cages of the US government, though. In November 2002, he was formally indicted by the US Department of Justice for illegally accessing ninety-two different computers ranging from every branch of the US military to private machines.[60] His charges came with the potential for up to ten years in a federal prison and a $250,000 fine. The US government tried for a decade to have McKinnon extradited and charged, but he prevailed after the charges were dropped.[61]

If the projects McKinnon claimed to find indeed exist, documentation should be available. Although it is difficult to accept that our technology has secretly progressed far enough to put such advanced equipment in orbit, and McKinnon had the ability to access the systems he mentioned, sufficiently more proof is required to confirm the US government has created a new space force of non-terrestrial officers.

November 14, 2004

On November 14, 2004, five men of the US Navy witnessed the now infamous "Tic Tac Incident" that was later confirmed by the Navy to involve an unidentified aerial phenomenon like those mentioned in the 2023 congressional hearing.

According to radio chatter, Gary Voorhis, Jason Turner, P.J. Hughes, Ryan Weigelt, and Kevin Day observed a fleet of Tic-Tac-shaped UAPs during a training operation about 100 miles off the coast of Southern California. They managed to record a relatively low-quality black and white video of the shape moving at over 100 knots (115 mph) regardless of wind conditions. Weigelt mentioned seeing one such UAP dropping at incredible speed from 80,000 feet and coming to a halt just over the water without any visible propulsion systems.

According to instrument readings on the USS *Princeton*, the UAPs matched no known cross section and did not register on any of their Friend-or-Foe (IFF) systems. In 2019, Voorhis told *Popular Mechanics*, "I couldn't make out details, but they'd just be hovering there, then all of a sudden, in an instant,

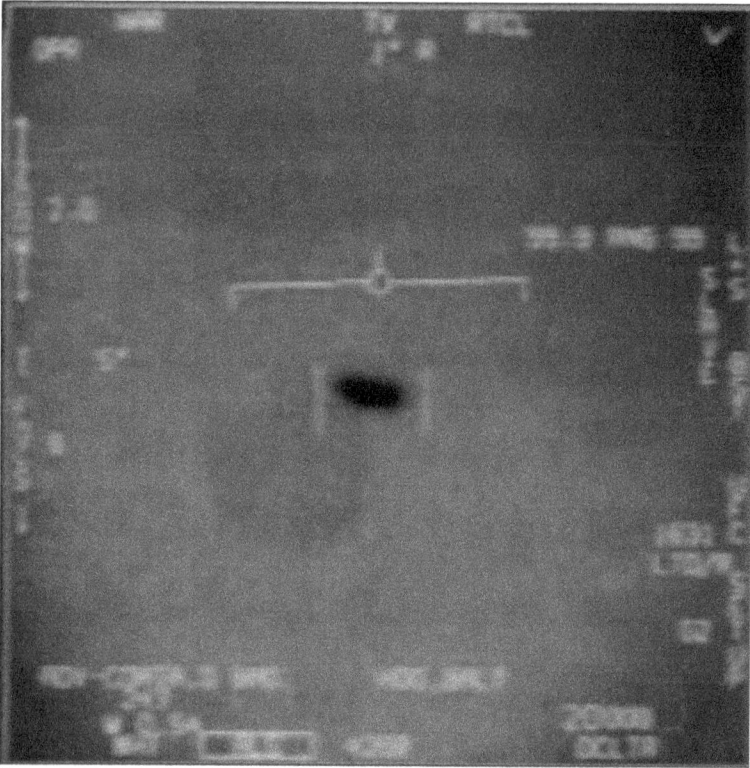

FIGURE 4.2: *Tic Tac video image.*

This screenshot captures the grainy black-and-white footage released by the US government of the Tic Tac-shaped unidentified aerial phenomenon observed near the USS *Princeton* during training exercises.

they'd dart off to another direction and stop again. At night, they'd give off a kind of a phosphorus glow and were a little easier to see than in the day."[62]

In an unofficial executive summary of the incident, the Department of Defense confirmed the descriptions of the UAPs as "elongated egg or Tic Tac shape with a discernable midline horizontal axis." The summary also included details on how two F/A-18Fs were diverted to observe and report the encounter. The F/A-18Fs that intercepted corroborated the description and speeds of the Tic Tac-shaped UAP[63].

Judging by the Navy's description, the F/A-18Fs were no match for the UAP they encountered. It seems to me the UAP visitors were observing the Navy's training operations, as they seem to have a pattern of doing when observed by military personnel going as far back as the Roswell incident near where nuclear testing was taking place or near Fort Dix and McGuire AFB.

December 16, 2017

Events similar to the Tic Tac incident eventually led to the formation of the Advanced Aerospace Threat Identification Program (AATIP) by the Pentagon. The program operated in secret from at least 2007 until 2017 when a US intelligence officer named Luis Elizondo stepped down from his position citing funding conflicts within the program. In an interview with *Politico,* Elizondo revealed that the program was potentially still operating, although he was unsure of its funding status.[64]

Programs such as the AATIP, operated by the Pentagon, routinely operate without congressional oversight and do so with checks written with US taxpayers' money. According to the *Politico* article, the AATIP alone ran with a budget of $22 million that Elizondo was aware of. Though the revelation of the program by the Pentagon showed that they were allegedly concerned that the observations by military personnel were sightings of secret foreign projects that may have been testing experimental propulsion systems and technology. While these UAP sightings are commonly attributed to extraterrestrials, it is entirely possible that world governments have experimental technologies capable of some of the incredible feats mentioned above.

Something else to consider when describing the physically impossible maneuvers being pulled off by these UAPs is that they are unmanned. Unmanned aerial vehicles, commonly referenced as drones, have been tested within the US military since the 1930s with radio-controlled torpedoes. It would be farfetched to believe human technology in 1947 was advanced enough to create the UAP recovered from Roswell, however, it is conceivable that scientists could be testing adapted technology as of 2004 and beyond after

FIGURE 4.3: *Saucer-type spaceship.*

having plenty of time to possibly reverse engineer what they found in 1947. If Twining's memo was truthful, it would have given the US government more than enough time to produce incredible prototypes straight out of science fiction. And if other governments had recovered their own technology from crashed UAPs, as the Pentagon fears, they could very well have their own prototypes darting through the skies. Such drones would be invaluable for intelligence collection around the world.

JUNE AND JULY 2020

In June 2020, the US Senate held hearings regarding budget appropriations for the 2021 fiscal year and mentioned the government's UAP Task Force. The task force, which at the time resided in the US Office of Naval Intelligence, is designed to standardize and collect reports on sightings at least twice a year. Since its inception, the task force has been practically plagued with reports of UAPs over many US sensitive military installations and cities.[65] The frequency and detailed descriptions of these reports make them impossible to ignore. They pose a potential threat to national security. These are also just the reports the US government is aware of and willing to reveal publicly. No doubt that there are similar task forces around the world collecting similar information that are either too secretive or too doubted to report publicly. Given many of these UAP reports are national security concerns, it is unlikely that opposing governments are willing to communicate about them for fear of revealing their potential weaknesses.

In *Popular Mechanics*, Andrew Daniels reported that Harry Reid, former Nevada Senator, believes that spacecraft from other worlds have crashed on our planet. He said material recovered from these crashes has been studied and used by government contractors for decades. This reinforces the claims of Colonel Corso, who contended that for two years he provided materials from the Roswell crash to contractors who used them in weapons and for technological development. Reid's statement is also in agreement with the deathbed declaration of Norma Gardner, who claimed to have cataloged over a thousand items taken from alien crashes.[66] Finally, it supports the deathbed declaration of former Lieutenant Haut, who confirmed the Roswell crash.

In July 2020, the *New York Times* reported that Eric Davis, the astrophysicist who consulted on the UFO project, said some of the materials that he had examined came from "off-world vehicles not made on this earth."[67]

5

Cosmic Ripple Effect

MILITARY ANALYSIS

Military analysis has developed over the years into a very sophisticated and comprehensive process. It is based on learning as much as possible about a potential adversary. Insight into the capabilities and probable actions of an adversary is the primary focus.

Why are They Here?

Despite popular culture, we cannot say with any certainty that extraterrestrials visiting Earth may be hostile. Based on the interactions humanity has experienced thus far, they seem to be more interested in observation than hostility. However, it is possible that one day humanity may come to blows with extraterrestrial visitors.

Beyond questionable reports from the 1968 Vietnam incidents, I have not found any information that shows conflict with UAPs. Even in the instance of the Vietnam incident, if true, the UAPs were returning the fire that was sent at them, not utilizing their own weapons against us. Yet that presents another problem entirely: we have no way of knowing how extensive their offensive capabilities are.

Knowledge of Senior Military Command

One of the most intriguing elements of military planning and analysis is the order of battle. This amounts to amassing information about an enemy's most senior command to predict how they will react in a particular situation. One great example of this goes back to the Civil War, the ancient proverb about "knowing your enemy," and General Lee's knowledge about General Grant. These generals knew each other before the war because they attended West Point together. During the war, after Grant experienced victories in the west, Confederate generals believed Grant would simply fade into the background after those successes. Lee, on the other hand, who personally knew Grant, told them Grant would *not* fade away. He would never quit.

As it turned out, Lee was correct. Even though it did not enable the South to win the Civil War, it illustrates the importance of knowing your opponents to better predict what they might do.

Supply and Communication

More issues essential to any military endeavor are supplies and logistics. Based on the sheer number of sightings, UAPs appear to supply themselves with anything they need. Perhaps they are secretly harvesting from our planet, or they are supplied from otherwise unseen larger vessels. This would be similar to the way fighter jets resupply on aircraft carriers.

Given repeated observations of UAPs moving with incredible speed and precision, sometimes in concert, it is no stretch to believe they have equally sophisticated communication methods.

If these other-world craft are so advanced, how have we allegedly recovered crashed UAPs? Without the revelations of top-secret studies, we can only assume that despite the advancement of their technology, a UAP is still a vehicle being piloted by a conscious being capable of error. Even the most foolproof technology can suffer from part failure and user error, after all.

It could be sheer dumb luck that UAPs have crashed and been recovered. For example, the lightning storms in the days leading up to the Roswell incident could have been what caused the "original" UFO to crash. Another possibility is that UAP visitors are here to secretly observe Earthlings, and some have fallen victim to other situations or natural dangers that we have adapted to, but they have not.

Everything we have observed about the duration of their visits and what they were doing indicates they were studying Earth and its inhabitants. Our developments into nuclear technology could have been of great interest to them, explaining why so many sightings have been reported near nuclear facilities. However, they could also be scouting locations of importance for future action.

With so many unknowns, nothing should be discounted as a possibility.

Weapons Capability

We can assume our conventional weapons would be all but useless against the UAPs we have witnessed to date. That would explain the push to develop higher tech weapons such as the Directed Energy Weapons the US Navy has experimentally begun using. If whistleblowers are to be trusted, it is possible the US government has been testing new technologies against recovered UAP materials in secret.

Although we have not been able to ascertain as much as we would like about the vulnerability of the extraterrestrials on board UAPs, we have some indicators. Roswell reports say that a soldier shot and killed one of the aliens with his M1 rifle. During the incident at McGuire Air Force Base, an alien was shot and killed with a .45 caliber handgun. This would surely indicate that when outside their craft, aliens are prone to the same dangers as humans are.

Interestingly, we also have not observed these beings wearing elaborate spacesuits but merely fabric garments—like the aliens at Roswell. We do not know if our planet's viruses or bacteria would affect them adversely.

FIGURE 5.1: *Gray Alien.*

Because we have observed at least three distinct vessel configurations, surely the military is considering the implications. Do these different types of spaceships come from the same source, or have we had visitors from multiple intelligent societies? If different alien groups are visiting us, the difficulty of our situation is compounded. Trying to determine the possible actions and strategy of one alien is hard enough.

It is certainly possible that the visits come from the same source and they have just developed different types of spaceships, the same way that we have developed different types of aircraft.

Conventional military analysis of possible alien intentions is hampered by the lack of information that we would otherwise possess about an adversary on Earth. One thing, however, is certain: the knowledge and technology necessary to travel great distances to visit Earth and to perform some of the feats we have seen show the superiority of their technology. Just how superior is the issue? We can only hope that any spacecraft analysis has provided some answers. Our ability to effectively deal with them is critical should they become hostile.

It is important that the Department of Defense has acknowledged that we have not only had encounters with UFOs but that we have been investigating this ongoing phenomenon. The challenge this could present to the world in the future may be like none other we have experienced.

TERRESTRIAL IMPACTS

When discussing the possibility that Earth has been visited by extraterrestrials and whether it has had an impact on humanity, it is impossible to discount the hundreds of reports of alien abductions. Thousands of people all around the world have claimed they were taken by extraterrestrials and subjected to physical examinations as well as experimentation. Some have even undergone hypnosis to help them recall their experiences that were seemingly blocked out or erased afterwards.

Betty and Barney Hill

One of the most famous reports of alien abduction is that of Betty and Barney Hill. On September 19, 1961, as the couple was driving home from their honeymoon in the White Mountains of New Hampshire, they felt they were being followed. According to their report, they spotted a UAP hovering above their vehicle. Suddenly, they were surrounded by a bright white light and blacked out. Two hours later, the couple regained consciousness and realized they were several miles down the road from where they had blacked out. After they got home, they noticed some marks on their car and that the zipper on Betty's dress was damaged.

The Hills kept the incident to themselves for quite some time. However, they were plagued by nightmares and decided to seek the help of a hypnotist to deal with their overall sense of anxiety. The hypnotist recorded their sessions during which they recalled suppressed memories of boarding a flying saucer and undergoing physical evaluations and experimentation. Betty especially seemed to be afraid of the experience and the beings performing the tests.[68]

Alien abduction stories like the Hills' are commonplace enough that some version of their story has become the basis for the alien abduction trope in media. No doubt readers will have seen a nearly exact recreation of the Hills' abduction story at some point.

My skepticism creeps in when considering how so many people have such similar, incredibly intimate experiences. Possibly, they are influenced by the media and stories shared around the world. That does not explain the Hills' incident, though, given they were reportedly the first to experience the phenomenon.

However, extraterrestrials, UFOs, and the framework of stories surrounding them were nearly as popular in the 1960s as they are in the 2020s. However, the number of reports expanded incredibly in those sixty years, but the stories remain the same.

So, why are these stories so prevalent? Are people making stories up to seek attention? Could it be related to mental illness?

Abduction Story Research

Richard McNally and Susan Clancy, both professors of psychology, sought an answer to the latter question. In a January 2024 article in *The Harvard Gazette*, William Cromie reported on their work. McNally and Clancy documented the story of Mark H., who believed he had been abducted by and had sex with extraterrestrials before being returned to his bed. In order to process his experience, Mark H. was placed under hypnosis and guided through an accurate recollection of his abduction, similar to how the Hills recalled their experiences.

As part of their research, McNally and Clancy performed the same study on a group of people who claimed to have been abducted and a group of people who underwent hypnosis to process some form of trauma not related to alien abductions. Interestingly, the results showed that the emotional responses to abduction were similar to, if not sometimes stronger than, those of the unrelated trauma victims.

Cromie shared McNally's explanation of their results:

The results underscore the power of emotional belief. People who sincerely believe they have been abducted by aliens show patterns of emotional and physiological response to these 'memories' that are strikingly similar to those of people who have been genuinely traumatized by combat or similar events.

McNally and Clancy also compared the alien abduction stories to those of sleep paralysis hallucinations, which are incredibly common around the world without any indication of mental illness. They also pointed out stories similar to the alien abductions that are related to mythical creatures such as devils and witches depending on the surrounding culture.[69]

In another *Harvard Gazette* article, Clancy explained, "Everyone in this group developed his or her belief of alien abduction after describing an episode that is consistent with sleep paralysis, a harmless but nonetheless frightening desynchronization of sleep cycles."[70]

Given the links, and the description of the drive, it is possible that the Hills had a similar experience while driving through the mountains and were fortunate enough not to coast off the road. While both reported separate yet related experiences, it is possible to form false memories and wholly believe them. It is also possible for one person's insistence to influence the mind of another to agree. Examples of sleep paralysis would also explain the all-too-common periods of blacking out and coming back to consciousness that are reported by abduction victims.

McNally and Clancy's study fails to account for reports of abduction that allegedly left tangible evidence behind. Unfortunately, their study was focused on the psychological aspects of abduction and memory formation.

Robert McCarty

Yet Roger McCarty from Ruby, Alaska, claimed he was tracked for more than two decades after being abducted in 2001. After being struck in the leg while hunting at night, he says he was implanted with a small device, and later abducted from his hunting cabin before being returned to his bed.

While working with the crew of the television series, *Aliens in Alaska*, McCarty had his leg scanned and found that the implanted leg was giving off radio waves. There was indeed a three-millimeter, pill-shaped object just under the skin of his leg.

Before Dr. Van Ravenswaay of Anchorage, Alaska, removed the device from McCarty's leg for analysis, he discovered the radio waves coming from that leg had stopped. There were no radio waves emitting from the object removed from his leg either.[71]

Unfortunately, beyond describing the object as a "cylinder of tissue," Dr. Ravenswaay and the production team of *Aliens in Alaska* did not elaborate further on the object or if it was sent out for additional testing. It would be enlightening to see what, if any, results came from testing the object removed from Mr. McCarty's leg; especially, given the claim that it was, at one point, emitting radio waves and McCarty's belief that he was possibly being tracked for so long.

Animal Abduction

Fortunately, human abductees did not undergo the treatment inflicted upon allegedly abducted animals. In the 1970s, there were several reports of cows in the western and midwestern US that were abducted and mutilated with astonishing precision.

Livia Gershon of *JSTOR Daily* wrote, "When local law enforcement agencies investigated, they frequently found that the cows' ears, eyes, rectums, and sex organs had been cut away with 'surgical precision.'"[72] Many of the animals' legs were broken as though they were dropped from a significant height, and, notably, typical carrion animals seemed to steer clear of the remains.

Locals seemed to blame Satanic Cults and UFOs for a while, but the popular opinion eventually settled on their cattle being mutilated in service of some secret government experimentation. Ranchers began firing on government helicopters flying over their lands, enough so that the Nebraska National Guard changed their operating altitude from 1,000 feet to 2,000 feet to escape potential gunfire.

Blaming cultists does not make much sense; they are unlikely to have the surgical skills to remove the animals' parts with such accuracy. Even if the US government was experimenting on these animals, the simple removal of parts and organs would not have prevented carrion animals from consuming the carcasses.

The FBI attempted to investigate the reports of cattle mutilation. However, outside of a few cases on Native American lands, they lacked the jurisdiction to do so. Their records show little more than headlines and research surrounding the reported mutilations.[73] Unfortunately, the lack of action or information from the US federal government does little to dissuade the idea that secret government experimentation was involved. If true, it would hardly be the first time the US government was involved in secret experimentation. The content of this book is plenty proof of that. Though it does beg the question of why they would have returned the cattle to their pastures when they were finished with them. Unless what happened after the cattle were returned to the fields was part of the experiment, I do not believe that to be the case.

Crop Circles

Another supposed mark of extraterrestrial visits that have interested the UFO community are crop circles. These intricate, often circular, patterns are embossed into crop fields all over the world, usually crops with lengthy stalks so the patterns stand out.

The conversation surrounding crop circles is a constant back and forth between those who believe the circles were placed by extraterrestrials and those who believe they were laid out by humans. In many cases, the circles seemed to appear overnight and were designed with great precision. The crops are not broken, either. They are carefully laid over in a way that appears to almost flow in the design's direction.

FIGURE 5.2: *Crop Circle, Aliens & UFO, June 16, 2023.*

The designs have been showing up as recently as 1976. Hundreds of circles appeared in South England near the famous Stonehenge site. They were later confirmed to be pranks laid by Doug Bower and Dave Chorley. The duo also claimed to have spawned several copycat designers that have been at work ever since.[74]

More recently, between 2005 and 2023, Wiltshire County, England, reported 380 crop circles. Despite the claim that crop circles are and have been pranks by clever artists, people such as the founder of the Crop Circle Exhibition, Monique Klinkenbergh, claim that some of the circles were placed instantaneously, a feat clearly impossible for humans.[75]

There have been claims that the first crop circle was reported in 1678, called, "The Mowing Devil." However, that report does not match the documented design of a crop circle. Author and scientist Jim Schnabel pointed out that all reports of crop circles have shown that the crops were laid over another, similar to the plank-and-rope method used by Bower and Chorley.[76] In the legend of the Mowing Devil, the crops were burned and mowed to a perfectly even level the farmer claimed was impossible for man to achieve. Unfortunately, the Mowing Devil is more of a standout report of strange behavior than an early recorded crop circle.

Crop circle designs have grown increasingly complex over the years. At one point, Bower and Chorley decided to increase the complexity of their designs because they were being attributed to weather phenomena instead of extraterrestrial or supernatural origins.

As the designs have grown in complexity, so has the need for faster methods of laying the crops over as tradition demands. Dr. Richard Taylor, a physicist from the University of Oregon, believes the designs are something of a scientific art movement in which the artists are working in conjunction with the scientists to find the best possible ways to imprint their designs. In his research, he found that some of the designs had blown out the back of the stalks to lay them over. He also hypothesized that it would be possible to remove the superheating element of a microwave oven and use it on the stalks of crops to lay them over with tremendous speed.[77]

Unfortunately, until the artists of these crop circles, be they human or otherwise, are caught or come forward with their techniques, the debate will rage on. As technology develops, so will the methods of creating crop circles. At some point, human designs will be capable of what we imagine extraterrestrials are capable of. Artists wielding microwave heaters, guided by satellite imagery and tracking, will be capable of laying out their most intricate patterns in the span of a short night.

Those who find these patterns are already finding hidden meaning in them, attributing their positions and designs to binary code and the keylines of the Earth being communicated to the skies and back.

Learning from Mystery

Roswell Artifacts

According to his book, *The Day After Roswell*, Corso heard a great many tales about Roswell and the highly sensitive materials uncovered at the crash site. While he was not present for the event itself, the Pentagon gave him access to materials allegedly from the Roswell incident including a locked filing cabinet in the corner of General Trudeau's office.

Trudeau told Corso that he was responsible for overseeing and utilizing the information in the cabinet. After that meeting, the cabinet was moved into Corso's office. Among the items in the cabinet, Corso found an autopsy report that provided excruciating detail about the study performed on the extraterrestrial bodies retrieved from the Roswell crash. The autopsy reports appeared to be medical reports that broke down the physiology and anatomical structure of the bodies.[78]

The bodies were described as approximately four-and-a half-feet tall with large heads and eyes, and gray-toned skin with a texture similar to that of a dolphin. Their internal organs differed from humans, and their bones were stronger and much more flexible. The descriptions, in medical terms, of the structure of these gray humanoids was very detailed.[79]

Among the items in the cabinet, Corso found the following:

- Cloth of a material unknown to Corso.

- A device that looked like a modern handheld laser pointer.

- A metallic band that may have been a headband for the extraterrestrials.

- Filaments similar to contemporary fiber optic strands.

- A piece of technology resembling a circuit board from a modern-day computer.

- A thin film that, when held up to the eye, provided daytime vision in the dark.[80]

After reviewing the above contents, it did not take Corso long to determine, along with documentation, that they were remnants of the Roswell crash site that had been preserved and safeguarded for over fourteen years. The general later informed Corso that he was tasked with developing a plan to utilize the information and materials in the cabinet. Corso was given free rein to work in secret with contractors to reverse engineer the materials without raising suspicions of where the materials came from in order to develop new technologies that would be useful to the military. There was significant concern surrounding the materials or the reverse-engineered technology getting into the hands of foreign adversaries.

Reverse Engineering

Over the next two years, Corso began working with US government contractors to find those most capable of reverse engineering the items in the cabinet. At the time, the US had existing contracts with companies such as Bell Labs and International Business Machines (IBM); both of whom were provided with the circuitry and thin glass filaments. According to Corso, the time spent reverse engineering the items in the cabinet either spearheaded the development of or accelerated the development of the following technologies:

- Integrated circuits, otherwise known as the microchip.
- Kevlar vests, allegedly sourced from the fabric on the extraterrestrials' bodies.
- Focused Energy or "Directed Energy" Weapons. DEWs, such as the AN/SEQ-3 Laser Weapon System, began being used as US Navy missile defense systems in 2014.
- Lasers, likely not the focused energy used in weapons, but the data transmission or reading methods.
- Night-vision technology, probably from the thin film Corso found.
- Fiber optic technology, used by the world as of 2024 for its improved internet speeds.
- Mind-control devices.

And those only address the items Corso listed in his book.

The headband, which reportedly caused strange mental reactions in those who put it on, could have been used to develop similar technologies. For example, the US Air Force is using cutting edge technology in helmets for F-35 pilots. These helmets allow the pilots a full 360-degree view of their surroundings while in flight.[81] The helmet also provides full day and night vision capabilities and allows them to control the jet through its heads-up display. Similar technology is used in the AH-64 Apache helicopter, whose helmet allows pilots to have similar heads-up displays as well as full control of the nose gun. The Apache nose gun can follow the gunner's line of sight, matching angle and direction for accurate shots on target. In the 1960s, these helmets, which represent just the publicly revealed examples of contemporary technology, would have once been considered in the realm of science fiction and nowhere close to a possible reality.

Defying Gravity

One of the areas of research stemming from materials collected from Roswell was creating aircraft that were capable of defying gravity. These experiments were similar to those supposedly being performed in Germany toward the end of World War II known as *die Glocke* or "the bell."[82] The idea was that the crashed Roswell craft had operated by creating its own gravitational pocket, allowing it to accelerate and change direction in ways that defied Earth's gravity. Such a method would also explain the pilots' ability to stay alive during such maneuvers, provided they had similar physiological constraints as humans.

If Nick Cook, the author of *The Hunt for Zero Point*, is to be believed, die Glocke would have been capable of amazing feats of propulsion or potentially even time travel. However, it may as well be Germany's own extension of the Roswell incident for all those that have tried to prove it. Interestingly, Cook alleges the German technology was given to the US government as part of a pardon deal, and that in the 1960s, the US may have begun work on such a project. It is possible the contents of Corso's cabinet were parts taken from scavenged Nazi WWII experiments. It is also possible that the US government and Nazi Germany both had possession of crashed extraterrestrial materials to experiment on.

General Trudeau clearly instructed Corso to reverse engineer the Roswell technology in order to advance US technological development of weapons and defenses. The end goal was not just to defend the US against foreign enemies, but potentially extraterrestrial enemies as well. Such caution is just as prudent in the 2020s as it was in the 1960s. With the capabilities humankind has observed so far, were the UAPs to attack us, we would have little to no defensive or offensive measures to counter them. Fortunately, so far, UAPs seem only concerned with observation. Clearly, these visitors from outer space possess knowledge and technology far beyond our understanding—both in the past and in the future. While the lasers and DEWs have proven useful against contemporary weapons, it is possible they will be effective against extraterrestrial craft as well, should they become aggressive.

Circuitry and Silicon

The development of integrated circuits and the use of silicon in electronics jumpstarted humanity's technological advancement. In 1961, computers were little more than calculating machines that took up entire rooms, buildings even. Since its inception in 1975, Moore's Law, the certainty that the technological potential of microchips doubles every two years, has proven true.

Microchips have gotten smaller at nearly the same rate they have grown in power, providing the world with leaps and bounds of advancement in the last seventy-plus years. As of 2023, Intel and other companies are moving to glass substrate processors, which have the potential to accelerate computer processing power even more.[83]

Glass was previously a computing medium only considered in science fiction. We saw it in books and movies such as *Star Trek, Star Wars,* and *Minority Report.* If what Colonel Corso wrote was true, it is possible that those efforts could be traced back to the technology the US took from the Roswell crash site. And if the testimonies made during the 2023 UAP hearing were true, it could mean that the US has continued collecting technology from newer extraterrestrial craft.

Laser Technology

As far as lasers and Corso's instructions are concerned, the US went from relatively simple light-emitting lasers to those that can transmit data across millions of miles in a matter of minutes.

In December 2023, NASA scientists were able to transmit a fifteen-second video of a cat named Taters over nineteen million miles or thirty-one million kilometers. According to NASA, the video took 101 seconds to reach Earth at a rate of 267 megabits per second.[84] The speed is significant when considering 256 megabits per second is faster than many people's home internet speeds. Naturally, the video of Taters the cat featured him sitting on a scientist's couch, chasing a laser.

6

Exploring the Universe

Our understanding of the universe has expanded broadly in the centuries since Galileo began exploring the celestial bodies through his telescope. In 1961, the Russians sent Yuri Gagarin into Earth's orbit. The next year, 1962, NASA sent John H. Glenn into orbit for the US on the Mercury-Atlas 6 mission. On July 20, 1969, NASA put boots on the moon. In 1975, NASA landed an unmanned rover on Mars, resulting in the famous photographs of the "face on Mars."[85]

For years, people were convinced the face on Mars was proof of alien life, a sign left by some civilization meant for humanity to discover. It turned out to be just another rock formation mixed with a healthy dose of *apophenia*, the human nature to find patterns and faces in everything we see.

We've continued to advance since then.

VOYAGERS I AND II

With 1977 came the longest-running missions to date, Voyagers I and II, which were launched into the deepest reaches of space, eventually leaving the solar system we call home. In 1990, we launched the Hubble telescope into orbit, where it provided the best images of deep space that had been recorded to date.[86] Humanity's hunger to learn more about what surrounds us proving insatiable yet again.

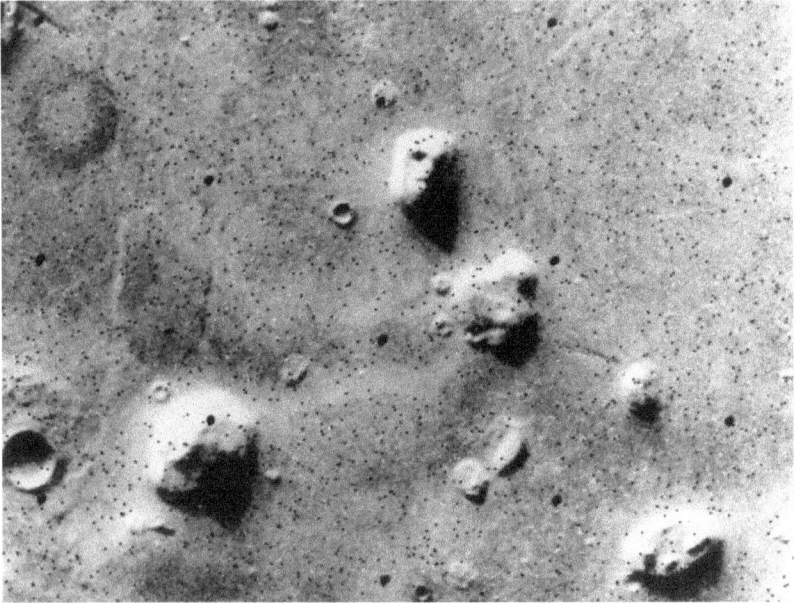

FIGURE 6.1: *The Face on Mars.*

Above is the picture of the Face on Mars near Cydonia, as taken by the Voyager I spacecraft in 1977. For years, this "face" convinced people that humans were being sent a message of some kind from the beings that must have once lived, or possible currently lived, on Mars.

FIGURE 6.2: *Galactic Wreckage in Stephen's Quintet.*

A group of five galaxies photographed in from the NASA/ESA Hubble Space Telescope.

YUTU-2 ROVER

In 2019, when the Chinese government sent a rover to the far side of the moon, the team discovered something glittering in the all-encompassing dust of the moon: glittering glass spheres. The mission scientists theorized the spheres were formed from the extreme heat caused by high velocity meteor impacts.[87]

In the same mission, the Yutu-2 deployed an advanced ground penetrating radar to further explore what was previously thought to be a giant ball of planetary offshoot and discovered massive lava tubes beneath the surface far larger than lava tubes found on Earth. The lava tubes are evidence of volcanic activity nearly one billion years ago and are large enough to potentially use as future lunar mission bases.[88] While the tubes are not a sign of alien activity, the radar scans proved the moon may not have always been a dead, dusty ball of rock and dust. It is also possible that extraterrestrials have or possibly would have used these tubes as bases or simply hiding areas from human observation. However, the study did not show any evidence in that regard.

PERSEVERANCE ROVER

At the time of this writing, NASA's most recent mission to Mars is the 2020 Perseverance Rover. On February 18, 2021, Perseverance landed on Mars and began collecting the highest definition photos and videos of Mars's surface that the world had seen. Unlike previous missions, Perseverance is capable of drilling into the surface and collecting samples for a future mission to potentially bring back to Earth for in-depth analysis. Perseverance even has the Mars Helicopter that it can launch for aerial surveys.

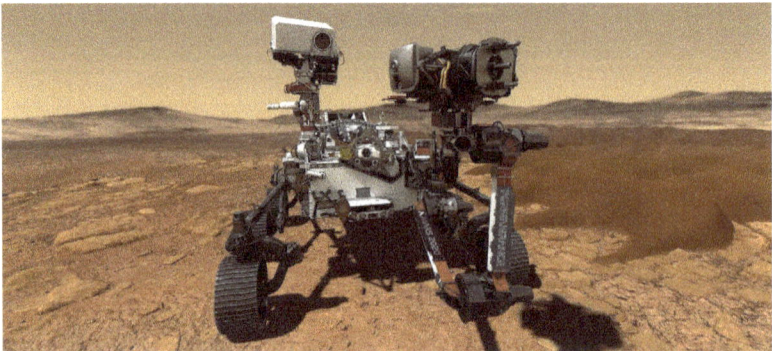

FIGURE 6.3: *Artist's Rendering of Perseverance Rover.*

Relative to the Martian background, Perseverance does not look all that big, but according to the mission website, it is ten feet long, nine feet wide, and seven feet tall; making it larger than most human beings. NASA hopes that Perseverance's survey capabilities will show signs of life on Mars within the Jezero Crater as well as test out technology they hope to use for crewed missions to Mars in the future.[89]

James Webb Space Telescope

In December 2021, NASA launched its most advanced orbital telescope yet, the James Webb Space Telescope (JWST) with the combined efforts of over 10,000 people in twenty years.[90] In the years since its deployment, JWST has shown us just how much detail we were missing from the Hubble images and allowed us to see deeper into space than ever before. The JWST is an infrared telescope intended to explore the deepest reaches of the universe, at least within the limitation of the speed of light. Light travels at roughly 186,000 miles per second, capable of circling the Earth at its equator 7.5 times per second. So, when looking at images of the famed Pillars of Creation, which are 6,500 light-years away from Earth, the information we can collect is 6,500 years old.

FIGURE 6.4: *Comparison of "Pillar of Creation" images as view through between Hubble telescope (left) and the James Webb Space Telescope (right).*

Above is a NASA comparison between the images of the Pillars of Creation taken by the Hubble and James Webb Space Telescope. Note that JWST was able to see more stars by orders of magnitude, both old and new stars still being formed in the past. It is likely that many of the stars are still coming together within the Pillars of Creation, given research suggests that stars take millions of years to form.

Nancy Grace Roman Space Telescope

NASA already has plans for the next, more specialized high powered space telescope: the Nancy Grace Roman Space Telescope (RST).[91] The RST is designed to use its wide field and coronagraph instruments to look as far away, and essentially back in time, where it may be possible to learn more about the creation of the universe, its components, and analyze exoplanets.

Beyond the Speed of Light

Until humanity reaches a level of technology capable of bypassing the physical boundaries of light speed, the most we can do is look out at the stars and attempt to explore the planets around us. Barring that, we can only hope to learn from an extraterrestrial species advanced enough to travel the stars. And, of course, we can question the possibility that Earth has already been visited by such beings.

The internet is rife with amateurs, articles, and alleged scientists claiming to have found evidence of extraterrestrial civilizations around the world. Unfortunately, discerning fact from fiction can be difficult on both sides of the exchange. It seems that every other week, we discover remnants of a lost civilization with levels of technology we did not believe existed in those eras.

Signs of Life in a Vast Universe

It is important to note that NASA officially states that there have been no provable signs of alien life found in space or on our home planet. As of October 2022, an article titled "About Astrobiology" by Marc Kaufman and posted on NASA's website, he wrote:

Ask most anyone whether life exists on other planets and moons, and the answer you'll get is a confident "yes!" Going back decades (and in many ways generations), we've been introduced to a menagerie of extraterrestrials good and bad. Their presence suffuses our entertainment and culture, and we humans seem to have an almost innate belief—or is it a hope—that we are not alone in the universe.

But that extraterrestrial presence on regular display is, of course, a fiction. No life beyond Earth has ever been found; there is no evidence that alien life has ever visited our planet. It's all a story.[92]

It is important to mention that Kaufman admits the fiction does not make the rule. Evidence of heat and water, the building blocks of life, have been found on Mars and Europa, neither of which have been explored thoroughly enough to write off the potential existence of life on those planets. Statistically, that could mean that life has existed on other distant planets that meet the conditions needed to facilitate life. The lack of observed evidence does not mean that alien life does not exist, merely that we have not yet found it.

Assuming we, the general public, listened to official sources around the world, the answer to "Are they out there?" would likely be a resounding "No." Or at least some variation on that. As noted earlier, NASA states that no definitive signs of extraterrestrials have yet been discovered. So, neither scientists nor governments will confirm their existence.

However, that does not mean an outright no should answer our titular question. Our observable universe is massive beyond understanding, and the limits of its possibilities have tickled the minds of scientists, observers, and science fiction writers alike for generations.

When I write that the universe is massive, many may picture terrestrial comparisons that can never do true justice to the universe's size. Nothing can, really. It's beyond imagination. Distance in the immensity of space is measured in light-years, or the amount of time in Earth years it takes for light to travel from one point to another. For comparison, the light from Earth's sun takes roughly eight minutes and twenty seconds to reach Earth and the sun is about 92.875 million miles (150 million kilometers) from Earth.[93] When distances begin to need light-years to be accurately measured, miles and kilometers are simply not up to the task.

By measuring the age of light reaching Earth, scientists at the European Space Agency were able to determine that our observable universe is nearly 28 billion light-years across when measuring Earth as the center of a sphere.[94] Even moving at the speed of light, capable of circumnavigating the Earth at its equator 7.5 times in a single second, it would take several times longer than life is estimated to have existed on Earth to cross even half of the known universe.

FIGURE 6.5: *Spiral Galaxy, NASA.*

Jan 09, 2014 · A photogenic and favorite target for amateur astronomers, the full beauty of nearby spiral galaxy M83 is unveiled in all its glory in this Hubble Space Telescope mosaic image. The vibrant magentas and blues reveal the galaxy is ablaze with star formation.

Bringing the scale of the conversation down several orders of magnitude, we return to a humble image of a single galaxy containing billions of stars, solar systems, and other celestial bodies. Messier 83 is a spiral galaxy similar to ours, the Milky Way Galaxy, which is relatively packed with what NASA has dubbed "exoplanets." Exoplanets are rocky planets close enough to stars to harbor life similar to life on Earth. As of March 2024, NASA had identified over 5,000 exoplanets capable of sustaining Earth-like life and have a long-running mission to try and explore those planets.[95] Much like the vastness of the universe, that number only covers the exoplanets that have been discovered.

THE PROBABILITY OF LIFE

According to NASA, the problem of searching for life on other planets is a mathematical one. Yes, there are thousands of potentially habitable planets that fall into the perfect zone around the right stars, but finding life on those planets is a shot in the dark. With the age of the universe and the incredible amount of exploration needed, the odds of finding both a habitable planet and one that bears life is already slim. Finding one such planet that has not only life, but sentient life capable of interstellar travel cranks that probability up even higher.

Yet the same math that makes intelligent life so difficult to find also all but guarantees its existence. NASA Researcher Ravi Kopparapu stated, "It's not a question of 'if,' it's a question of 'when' we find life on other planets. I'm sure in my lifetime, in our lifetime, we will know if there is life on other worlds."[96] That was in 2021.

Astronauts around the world are some of the brightest and most capable minds humanity has selected to explore space. Any NASA astronaut must have a master's degree in an applicable field, usually in science, technology, engineering, or math (STEM), and must complete the rigorous two-year astronaut training program in order to go to space. All in all, it takes around ten years of dedicated effort to become an astronaut with the qualifications to observe and make theories about the very space they have explored and studied.

When astronaut Buzz Aldrin was asked by *Forbes* if he believed life existed out in the expanse of the universe, he replied:

I'd have to go to Stephen Hawking and some people like that that think big. And when you think big, you find how big our Milky Way Galaxy is. One hundred thousand light-years. That's a billion stars, and one galaxy is of billions and billions of galaxies like our Milky Way. Surely there's some place sort of like Earth.[97]

Another NASA moonwalker, Ed Mitchell, tandem record holder for longest moonwalk, told the *Guardian*, "Oh yes. There's not much question at all that there is life throughout the universe. I'm totally sure that we are not alone."[98]

Yet another astronaut, this time Helen Sharman, the first Briton in space, had the following to say:

Aliens exist, there's no two ways about it. There are so many billions of stars out there in the universe that there must be all sorts of different forms of life. Will they be like you and me, made up of carbon and nitrogen? Maybe not. It's possible they're here right now and we simply can't see them.[99]

So, while humanity has not yet confirmed the existence of extraterrestrials, the probability of such life existing coupled with the age of the universe all but guarantees "they" are indeed out there. Assuming that the powers that be are to be trusted, and that is a big assumption given the topics we have covered thus far, why has humanity not been contacted by these advanced beings?

In 1950, physicist Enrico Fermi famously asked over lunch, "Where is everybody?" While the lunchtime conversation was not documented in detail, only recalled later by those who were present, the logic is broken down as follows. Given the billions of stars just in the Milky Way Galaxy and the high probability that many of those stars have Earth-like exoplanets in the habitable zone, there must be civilizations that have evolved and explored

FIGURE 6.6: *Messier 16 NASA*

Above: Messier 16, or the Eagle Nebula, also known as the Pillars of Creation, as captured by the Hubble Space Telescope. The Pillars of Creation are massive star creation clusters that span more than nine light-years. This amazing structure is 7,000 light-years away from Earth. In the updated version captured by the James Webb Space Telescope, it was revealed that the "stars" behind the Pillars are actually bright galaxies surrounded by millions of more distant galaxies.

to the point of being capable of interstellar travel. Yes, Earth and our sun are ancient in their origin, there are stars that are several billions of years older than it. More than enough time for civilizations to have risen to the

FIGURE 6.7: *Active quasar illustration, NASA.*

Above: An illustration of a galaxy with an active quasar created by NASA. Quasars are the brightly lit cores of galaxies where gas and dust swirl around the black hole at the center in astronomical quantities.

point of being able to visit. So, why is it that with all those factors playing in concert, we on Earth have yet to see any evidence of faster-than-light travel or extraterrestrial visits?

Of course, as mentioned earlier, the same math that makes it clear that we are not alone could be the very thing that guarantees we stay alone. Humanity has been trying for ages to reach the depths of space and contact other species in the cosmos to no avail. The universe is a mind-bogglingly large ocean to be searching for a specific kind of fish in. Other advanced civilizations could be having the same problems finding us as we are them.

WE NEED TO KNOW THE TRUTH

The assumption that Earth has not been visited by extraterrestrials capable of interstellar travel also hinges on the truth not being covered up. If the mysteries surrounding Roswell and countless other unidentified aerial phenomena interactions have proven, the world governments are no strangers to hiding the truth from us. Whether for the benefit of the general public or themselves as controlling organizations, without their disclosure of events, we may never know if Earth has been visited or communicated with an advanced alien civilization.

Are the world powers so concerned for our safety or their own self-interests that they would be willing to hide the truth from the public? If Corso's book is to be believed, based on the economic titans in the technology sector, there may be a massive interest in monetary or political gain from hiding the truth.

On a more depressing note, it is possible that we have been visited and are little more than a rural pit stop, or nature preserve, for extraterrestrials to swing by and visit without interacting with us. That would certainly fall in line with the pattern of observation and lack of interaction found in alien encounter reports, both civilian and military. I can imagine that our technological and societal progress would be of immense value to an advanced sentient species exploring the cosmos.

Humankind already does the same with evolving species on Earth. Think of the National Geographic documentarians that spend their entire lives and careers exploring the wilderness just to capture the briefest moments of natural wonder without the influence of humankind. Their efforts may not always work, as the photographer's mere presence is enough to change a handful of imperceptible factors in their subject's environment; but the moments they capture and report are often genuine enough to change how we believe those creatures live. If the UAP reports so far are considered in this context, we as humans have certainly seen our visitors and been greatly affected by their presence in several ways they may not have intended.

A simpler explanation for why we may have not been visited by extra-terrestrials is the cost of doing so. Not necessarily monetary cost, though that is a massive factor for humankind, but fuel and technology. As we have already covered, the known universe is staggeringly immense and traversing it at light speed would take billions on billions of years. More than likely longer than most civilizations or species would survive, certainly longer than many species on Earth have survived and evolved. It could be that we simply have not been reached yet, assuming Earth has been discovered by these hypothetical advanced beings. Faster-than-light travel is theoretically possible, but at massive energy costs and technological capabilities well beyond humanity's as of 2024.

Physical Matter and the Speed of Light

The trick with faster-than-light travel is that physical matter as we know it is incapable of traveling at such speeds. "Blistering speed" does not do the idea of light speed justice. Humans can barely handle the acceleration of high-speed jets and will pass out or outright die when pulling too many G-forces that halt the blood from pumping through our bodies to keep us alive. Speeds even nearing light speed will likely reduce human bodies to atoms and string us out behind. Accelerating up to light speed at a reasonable pace would do exactly that. At a "healthy" speed of a freefall, getting up to light speed would take nearly one year provided the pilot was in space and had a clear eleven-month area to speed up in. However, as physicist Michael Pravica pointed out, as an object's speed increases, so does its mass, meaning that object suddenly takes significantly more energy to move than it did at the outset.[100]

In an effort to circumvent that wonderful twist in physics, a Mexican physicist by the name of Miguel Alcubierre devised the Alcubierre Drive. Alcubierre's drive was designed to stretch time and space in front and behind it, forming a sort of bubble that allowed space to move around the craft instead of the craft moving through space. Because the craft would not be moving through space, but the other way around, it circumvents the restrictions of speed, mass, and acceleration that stop human beings from moving at faster-than-light speeds.[101]

The problem with the Alcubierre Drive is that humankind, even more than thirty years later, has yet to sort out the energy cost of making the Alcubierre Drive work on a functional scale. They also face the practical issues of being able to create and subsequently escape the bubble in space, keeping the drive a theory for the time being.

FIGURE 6.8: *Artist rendering of the Alcubierre Drive in action, NASA.*

WORMHOLES

Following the trend of the Alcubierre Drive is another method humanity might traverse the cosmos unhindered by the regular laws of physics. Albert Einstein once theorized the existence of wormholes as "white holes" that were the antithesis of black holes. Wormholes are theorized to be spherical pockets in space that are connected by a tube that runs between space and time.[102]

Ideally, humankind would be able to traverse these wormholes, bridging the gap between massive distances that would take generations to travel at reasonable speeds. However, while Einstein's theory and black holes exist, we have yet to actually find a wormhole. Given they would likely look similar to a bending of visual space in the middle of nothingness, it's more likely we will find one by accident than by combing the heavens inch by inch. We would also need to be able to reach the wormhole and figure out how to safely traverse it. Without knowing what was on the other side, we could be sending astronauts to their death.

There is also the issue of communication across such massive distances. If humanity were to live across the universe's great expanse, we would need the capability to send and receive data across light-years. Radio waves travel at the speed of light, and we have figured out to beam information over great distances via lasers—such as the cat video mentioned earlier. Yet even those methods would take years to send and receive communication across the scale of the universe. This is not feasible for normal operations.

Quantum Entanglement

The European Space Agency (ESA) is researching quantum entanglement communications. Quantum entanglement acknowledges that particles across long distances can share a spatial proximity that could theoretically be used to transmit data.[103]

The data, radio waves in this example, would interact with one particle and be transmitted through the other particle despite how far away it may be. The process is similar to the theory behind traveling through a wormhole, though on a much smaller scale. According to ESA researchers, quantum entanglement communications would enable secure transmissions since they would not be intercepted between the two particles; at least not by any methods we are aware of.

Naturally, these ideas are still in their infancy as of 2024 and have not yet been proven to work as researchers hope.

If humanity is facing technological difficulties such as those surrounding the Alcubierre Drive, it is possible that other intelligent species in the universe are struggling with the same issues. That would go a long way toward explaining why we have been unable to observe them in any capacity, again assuming the world governments are not hiding the fact that we already have met extraterrestrials.

FIGURE 6.9: *Carina Nebula, NASA.*

THE GREAT FILTER

In 1998, Robin Hanson, an associate professor of economics at the Future of Humanity Institute of Oxford University, discussed the idea of the Great Filter. The Great Filter explains Hanson's view of why the Earth has likely not been contacted by or colonized by extraterrestrials. By Hanson's logic, all known life reaches a point when it begins to expand and colonize areas beyond its origin. Humans have been doing just that on Earth for most of written history. We may be reaching a point where we begin to expand beyond Earth and into the stars to harvest, at a distance, the resources and acreage needed to sustain human exponential growth and expansion. Hanson wrote:

Within the next million years (at most) therefore, our descendants seem to have a foreseeable (greater than one in a thousand) chance of reaching an "explosive" point, where they expand outward at near the speed of light to colonize our galaxy, and then the universe, easily overpowering any less developed life in the way. FTL (faster than light) travel would imply even faster expansion.[104]

So, based on that expectation of life, why have we not been contacted or colonized by a more advanced civilization? Given the age of the universe and the math discussed earlier in this chapter, it stands to reason that someone out there would have reached the level of expansion Hanson wrote about. Even on Earth, there exists evidence of a natural progression and cycle of cataclysmic change just short of a billion years per cycle, but some coming as shortly as half of that.

Hanson believed that other civilizations in the universe may be under similar cyclical restraints that keep them from reaching the stars. Such a cycle could even be by design, if you believe in some greater power, to keep civilizations below a certain technological threshold from expanding into the stars.

However, humankind may be pushing that boundary sooner than later. If we were to follow Hanson's estimation that humans will be colonizing beyond our solar system within a million years, that would put us out of reach of most cataclysmic events. His theory of a great filter estimates that if humankind can move beyond the range of an exploding star, we will be at a point where our expansion cannot be stopped.

Yes, humankind would be fractured at that point, due to the unfathomable distances that would exist between us, but we would be vastly more advanced than our origins and be equipped to continue advancing.

FIGURE 6.10: *Monocerotis V838, NASA.*

Above: The V838 Mon light echo, which continues to spread through space and provide a unique view of the thinned areas every time Hubble or JWST captures an image of it.

FIGURE 6.11: *Veil Nebula, NASA.*

Above: The Veil Nebula supernova remnant. The ever-spreading remnants of a star that collapsed over 8,000 years ago.

FIGURE 6.12: *The James Webb Space Telescope (JWST).*

FIGURE 6.13: *The Hubble Space Telescope.*

ARE THEY OUT THERE?

I leave you with this question and my answer. I have provided some of the best evidence I can find to support all sides of the question: yes, no, maybe? Alongside that information, I have shown that world governments have long histories of hiding the truth from us, and people are hungrier than ever to have that information revealed in full. We have a natural drive to expand and explore, and hiding the truth does little more than hamper that drive, and it needs to stop.

Even if the governments of the world were to be taken at face value that Earth has not been visited by extraterrestrials, the math is right there. Physicist and writer Neil deGrasse Tyson once said, "Math is the language of the universe. So, the more equations you know, the more you can converse with the cosmos." We can trust the math. I trust the math.

Even if we have not been visited, there is no way we are alone in this universe. So yes, they are out there.

APPENDIX A

Haut's Deathbed Confession

2002 SEALED AFFIDAVIT OF
WALTER G. HAUT
DATE: December 26, 2002, WITNESS: Chris Xxxxx
NOTARY: Beverley Morgan

(1) My name is Walter G. Haut.

(2) I was born on June 2, 1922.

(3) My address is 1405 W. 7th Street, Roswell, NM 88203

(4) I am retired.

(5) In July 1947, I was stationed at the Roswell Army Air Base in Roswell, New Mexico, serving as the base Public Information Officer. I had spent the 4th of July weekend (Saturday, the 5th, and Sunday, the 6th) at my private residence about ten miles north of the base, which was located south of town.

(6) I was aware that someone had reported the remains of a downed vehicle by midmorning after my return to duty at the base on Monday, July 7. I was aware that Major Jesse A. Marcel, head of intelligence, was sent by the base commander, Col. William Blanchard, to investigate.

(7) By late in the afternoon that same day, I would learn that additional civilian reports came in regarding a second site just north of Roswell. I would spend the better part of the day attending to my regular duties hearing little if anything more.

(8) On Tuesday morning, July 8, I would attend the regularly scheduled staff meeting at 7:30 a.m. Besides Blanchard, Marcel; CIC Capt. Sheridan Cavitt; Col. James I. Hopkins, the operations officer; Major Patrick Saunders, the base adjutant; Major Isadore Brown, the personnel officer; Lt. Col. Ulysses S. Nero, the supply officer; and from Carswell AAF in Fort Worth, Texas, Blanchard's boss, Brig. Gen. Roger Ramey and his chief of staff, Col. Thomas J. DuBose were also in attendance. The main topic of discussion was reported by Marcel and Cavitt regarding an extensive debris field in Lincoln County approx. seventy-five miles NW of Roswell. A preliminary briefing was provided by Blanchard about the second site approx. forty miles north of town. Samples of wreckage were passed around the table. It was unlike any material I had or have ever seen in my life. Pieces, which resembled metal foil, paper thin yet extremely strong, and pieces with unusual markings along their length were handled from man to man, each voicing their opinion. No one was able to identify the crash debris.

One of the main concerns discussed at the meeting was whether we should go public or not with the discovery. Gen. Ramey proposed a plan, which I believe originated with his bosses at the Pentagon. Attention needed to be diverted from the more important site north of town by acknowledging the other location. Too many civilians were already involved, and the press already was informed. I was not completely informed how this would be accomplished.

(10) At approximately 9:30 a.m., Col. Blanchard phoned my office and dictated the press release of having in our possession a flying disc, coming from a ranch northwest of Roswell, and Marcel flying the material to higher headquarters. I was to deliver the news release to radio stations KGFL and KSWS, and newspapers the Daily Record and the Morning Dispatch.

(11) By the time the news had hit the wire services, my office was

inundated with phone calls from around the world. Messages stacked up on my desk, and rather than deal with the media concern, Col. Blanchard suggested that I go home and "hide out."

(12) Before leaving the base, Col. Blanchard took me personally to Building 84, a B-29 hangar located on the east side of the tarmac. Upon first approaching the building, I observed that it was under heavy guard both outside and inside. Once inside, I was permitted from a safe distance to first observe the object just recovered north of town. It was approx. twelve to fifteen feet in length, not quite as wide, about six feet high, and more of an egg shape. Lighting was poor, but its surface did appear metallic. No windows, portholes, wings, tail section, or landing gear were visible.

(13) Also from a distance, I was able to see a couple of bodies under a canvas tarpaulin. Only the heads extended beyond the covering, and I was not able to make out any features. The heads did appear larger than normal and the contour of the canvas over the bodies suggested the size of a ten-year-old child. At a later date, in Blanchard's office, he would extend his arm about four feet above the floor to indicate the height.

(14) I was informed of a temporary morgue set up to accommodate the recovered bodies.

(15) I was informed that the wreckage was not "hot" [radioactive].

(16) Upon his return from Fort Worth, Major Marcel described to me taking pieces of the wreckage to Gen. Ramey's office and after returning from a map room, finding the remains of a weather balloon and radar kite substituted while he was out of the room. Marcel was very upset over this situation. We would not discuss it again.

(17) I would be allowed to make at least one visit to one of the recovery sites during the military cleanup. I would return to the base with some of the wreckage which I would display in my office.

(18) I was aware two separate teams would return to each site months later for periodic searches for any remaining evidence.

(19) I am convinced that what I personally observed was some type of craft and its crew from outer space.

(20) I have not been paid nor given anything of value to make this statement, and it is the truth to the best of my recollection.

THIS STATEMENT IS TO REMAIN SEALED AND SECURED UNTIL THE TIME OF MY DEATH, AT WHICH TIME MY SURVIVING FAMILY WILL DETERMINE ITS DISPOSITION.

Signed: Walter G. Haut
Signature Witnessed by: Chris Xxxxxx
Dated: December 26, 2002
Source: Tom Carey & Donald Schmitt, Witness to Roswell, 2007

APPENDIX B

The Twining Memo

SUBJECT: AMC Opinion Concerning "Flying Discs"
TO: Commanding General

Army Air Force

Washington, DC.

ATTENTION: Brig. General George Schulgen

AC/AS-2

1. As requested by AC/AS-2 there is presented below the considered opinion of this command concerning the so-called "Flying Discs." This opinion is based on interrogation report data furnished by AC/AS-2 and preliminary studies by personnel of T-2 and Aircraft Laboratory, Engineering Division T-3. This opinion was arrived at in a conference between personnel from the Air Institute of Technology, Intelligence T-2, Office, Chief of Engineering Division, and the Aircraft, Power Plant and Propeller Laboratories of Engineering Division T-3.

2. It is the opinion that:

a. The phenomenon is something real and not visionary or fictitious.

b. There are objects probably approximating the shape of a disc, of such appreciable size as to appear to be as large as man-made aircraft.

c. There is a possibility that some of the incidents may be caused by natural phenomena, such as meteors.

d. The reported operating characteristics, such as extreme rates of climb, maneuverability (particularly in roll), and motion which must be considered evasive when sighted or contacted by friendly aircraft and radar, lend belief to the possibility that some of the objects are controlled either manually, automatically, or remotely.

e. The apparent common description is as follows:

(1) Metallic or light-reflecting surface.

(2) Absence of trail, except in a few instances where the object apparently was operating under high performance conditions.

(3) Circular or elliptical in shape, flat on bottom, and domed on top.

(4) Several reports of well-kept formation flights varying from three to nine objects.

(5) Normally, no associated sound, except in three instances a substantial rumbling roar was noted.

(6) Level flight speeds normally above 300 knots are estimated.

f. It is possible within the present US knowledge—provided extensive detailed development is undertaken—to construct a piloted aircraft which has the general description of the object in sub-paragraph (e) above which would be capable of an approximate range of 7,000 miles at subsonic speeds.

g. Any development in this country along the lines indicated would be extremely expensive, time consuming, and at the considerable expense of current projects, and therefore, if

directed, should be set up independently of existing projects.

h. Due consideration must be given the following:

(1) The possibility that these objects are of domestic origin - the product of some high security project not known to AC/AS-2 or this Command.

(2) The lack of physical evidence in the shape of crash recovered exhibits which would undeniably prove the existence of these subjects.

(3) The possibility that some foreign nation has a form of propulsion

possibly nuclear, which is outside of our domestic knowledge.

3. It is recommended that:

a. Headquarters, Army Air Forces, issue a directive assigning a priority, security classification and code name for a detailed study of this matter to include the preparation of complete sets of all available and pertinent data which will then be made available to the Army, Navy, Atomic Energy Commission, JRDB, the Air Force Scientific Advisory Group, NACA, and the RAND and NEPA projects for comments and recommendations, with a preliminary report to be forwarded within fifteen days of receipt of the data and a detailed report thereafter every thirty days as the investigation develops. A complete interchange of data should be affected.

4. Awaiting a specific directive AMC will continue the investigation within its current resources in order to more closely define the nature of the phenomenon. Detailed Essential Elements of Information will be formulated immediately for transmittal through channels.

Image Credits
and Permissions

The author gratefully acknowledges the following sources for images. Those images without noted permissions for reproduction are in the public domain.

CHAPTER 1

FIGURE 1.1 pg. 5	"Daniel Potter, Assistant Curator with the National Museum of Scotland, with a casing stone from the Pyramids of Giza," by Gary Shaw. Used with permission of *Apollo The International Art Magazine*, National Museums Scotland. https://www.apollo-magazine.com/egypt-questions-national-museum-of-scotlands-right-to-giza-pyramid-stone/
FIGURE 1.2 pg. 6	"Rendering of Pyramids of Giza, Newly Built," used courtesy of creator, Joshua Smith.
FIGURE 1.3 pg. 6	"Pyramids of Giza as they are Today," used under the content license of Pixabay.com, https://pixabay.com/service/license-summary/. https://pixabay.com/photos/pyramids-egypt-giza-archeology-2159286/

FIGURE 1.4 pg. 9	"Elongated Skull," used with license from Getty Images.
FIGURE 1.5 pg. 11	"Map of Göbekli Tepe," in the public domain. Obtained from the Smithsonian Museum. https://th-thumbnailer.cdn-si-edu.com/5TbtSZ1NQm6xbiF674OG-PVaYPrA=/fit-in/1072x0/https://tf-cmsv2-smithsonianmag-media.s3.amazonaws.com/filer/gobeklitepe_nov08_11.jpg.
FIGURE 1.6 pg. 12	"Gobleki Tepe Megalith Ring," by Teomancimit, Creative Commons license https://creativecommons.org/licenses/by-sa/3.0, via Wikipedia. https://upload.wikimedia.org/wikipedia/commons/thumb/d/d5/G%C3%B6bekli_Tepe%2C_Urfa.jpg/1280px-G%C3%B6bekli_Tepe%2C_Urfa.jpg.
FIGURE 1.7 pg. 13	"Carving on T-shaped Stone," Used with license from pixabay.com. https://pixabay.com/photos/g%C3%B6beklitepe-gobekli-tepe-urfa-6126904/
FIGURE 1.8 pg. 15	"Inga Stone from the distance," by Lucia Barreiros da Silva, Creative Commons license, https://creativecommons.org/licenses/by-sa/3.0, via Wikipedia. https://en.wikipedia.org/wiki/Ing%C3%A1_Stone#/media/File:Itacoatiaras_de_Ing%C3%A1_-_Ing%C3%A1_Para%C3%ADba_Brasil.jpg
FIGURE 1.9 pg. 15	"Inga Stone carvings in detail," by Zelma Brito, Creative Commons license, https://creativecommons.org/licenses/by-sa/3.0, via Wikipedia. https://en.wikipedia.org/wiki/Ing%C3%A1_Stone#/media/File:Pedra_do_Ing%C3%A1_-_Para%C3%ADba_-_Brasil_-_panoramio.jpg
FIGURE 1.10 pg. 16	"The London Hammer in stone," by S. J. Miba, Creative Commons license, https://creativecommons.org/licenses/by-sa/3.0, via Wikipedia. https://en.wikipedia.org/wiki/London_Hammer#/media/File:London_Hammer.jpg

Figure 1.11 pg. 17	"Fake Astronaut Carving," used with license from Getty Images.

Chapter 3

Figure 3.1 pg. 35	Front page of the *Roswell Daily Record,*" on July 8, 1947, Public Domain, via Wikimedia Commons. https://commons.wikimedia.org/wiki/File:RoswellDailyRecordJuly8,1947.jpg
Figure 3.2 pg. 40	"Portrait of Brigadier General George Schulgen," courtesy of the US Air Force, Public Fair Use image. https://www.af.mil/About-Us/Biographies/Display/Article/2995977/george-francis-schulgen/.
Figure 3.3 pg. 41	"Portrait of Lieutenant General Nathan Twining," courtesy of the National World War II Museum, Public Fair Use image. https://www.ww2online.org/image/lieutenant-general-nathan-f-twining-1945
Figure 3.4 pg. 43	"Page from the FBI report on the MJ-12," courtesy of the FBI Records: The Vault. Public fair use image.
Figure 3.5 pg. 45	Letter sent to Secretary Forrestal by President Truman, September, 1947," unknown origin, Public Fair Use image.

Chapter 4

Figure 4.1 pg. 58	"Rendering of a Triangular Spaceship," used with license from Getty Images.
Figure 4.2 pg. 60	"Still from Pentagon release of 'Tic Tac Incident' FLIR footage," captured by USG FLIR cameras, Public Fair Use image.
Figure 4.3 pg. 62	"Rendering of a Saucer-style Spaceship," used with license from Getty Images.

CHAPTER 5

FIGURE 5.1 pg. 68	"Rendering of a Gray Alien," used with license from Getty Images.
FIGURE 5.2 pg. 74	"Photograph of Crop Circles in Switzerland," by Jabberocky, Public Domain, Wikimedia. https://en.m.wikipedia.org/wiki/File:CropCircleW.jpg

CHAPTER 6

FIGURE 6.1 pg. 82	"Satellite image of The Face on Mars," NASA, Public Domain. https://cdn.mos.cms.futurecdn.net/BrEpBuisVMDQz3p6WuBKAR.jpg
FIGURE 6.2 pg. 83	"Galactic Wreckage in Stephen's Quintet," NASA, Public Domain. https://esahubble.org/images/heic0910i/
FIGURE 6.3 pg. 84	"Artist Rendering of Perseverance Rover," NASA, Public Domain. https://www.nasa.gov/wp-content/uploads/2023/02/edu_srch_meet-per-severance.png
FIGURE 6.4 pg. 85	"Pillars of Creation Hubble vs. JWST Comparison," NASA, Public Domain. https://www.nasa.gov/wp-content/uploads/2022/10/stsci-01gfn-r1kzzp67ffgv8y26kr0vw.png
FIGURE 6.5 pg. 88	"Spiral Galaxy M83," captured by the Hubble Space Telescope, NASA, Public Domain. https://science.nasa.gov/wp-content/uploads/2023/04/hs-2014-04-a-xlarge_web_0-jpg.webp?w=1024&format=webp
FIGURE 6.6 pg. 91	"Messier 16," captured by the Hubble Space Telescope, NASA, Public Domain. https://science.nasa.gov/wp-content/uploads/2023/04/heic1501b-jpg.webp?w=1024&format=webp

FIGURE 6.7 *pg. 92*	"Illustration of an Active Quasar," NASA, Public Domain. https://science.nasa.gov/wp-content/uploads/2023/04/stsci-h-2010a-d-1280x720-1.png?w=1024&format=webp
FIGURE 6.8 *pg. 95*	"Artist Rendering of Alcubierre Drive," NASA, Public Domain. https://en.m.wikipedia.org/wiki/File:Wormhole_travel_as_envisioned_by_Les_Bossinas_for_NASA.jpg
FIGURE 6.9 *pg. 97*	"The Carina Nebula," captured by the James Webb Space Telescope, NASA, Public Domain. https://www.nasa.gov/wp-content/uploads/2023/03/main_image_star-forming_region_carina_nircam_final-5mb.jpg
FIGURE 6.10 *pg. 99*	"Monocerotis V838," NASA, Public Domain. https://science.nasa.gov/image-detail/idl-tiff-file-42/
FIGURE 6.11 *pg. 100*	"The Veil Nebula," captured by the Hubble Space Telescope, NASA, Public Domain. https://science.nasa.gov/wp-content/uploads/2021/04/veil-nebula-potw2113a-jpg.webp?w=1024&format=webp
FIGURE 6.12 *pg. 100*	"The James Webb Telescope," NASA, Goddard Space Flight Center, Conceptual Image Lab, Adriana Manrique Gutierrez, Public Domain. https://www.flickr.com/photos/nasawebbtelescope/51412207042/in/album-72157624413830771/
FIGURE 6.13 *pg. 101*	"The Hubble Telescope," NASA, Public Domain. https://science.nasa.gov/image-detail/hubble-space-telescope-hst/

Bibliography

1. Aldrin, Buzz. "Buzz Aldrin on the Existence of Extraterrestrials in Our Galaxy | Forbes." YouTube, October 23, 2021. https://www.youtube.com/watch?v=lNJpVDdUeUU.

2. Randle, Kevin. "A Little Help with Norma Gardner," February 22, 2016, http://kevinrandle.blogspot.com/2016/02/a-little-help-with-norma-gardner.html.

3. April Holloway. "Initial DNA Analysis of Paracas Elongated Skull Released – with Incredible Results." Ancient Origins *Unraveling the Mysterious of the Past*, September 28, 2022. https://www.ancient-origins.net/news-evolution-human-origins/initial-dna-analysis-paracas-elongated-skull-released-incredible.

4. Bain, Rachel. "Huge Hydrogen & Oxygen Balloon Explosion." UW-Madison Kaltura MediaSpace. Accessed September 8, 2024. https://mediaspace.wisc.edu/media/Huge+Hydrogen+%26+Oxygen+Balloon+Explosion/1_gjuirv0a.

5. BBC Staff. "Hacker Gary McKinnon Turns into a Search Expert." BBC News, July 28, 2014. https://www.bbc.com/news/technology-28524909.

6. Bender, Bryan. "The Pentagon's Secret Search for UFOs," *Politico Magazine*, December 16, 2017. https://www.politico.com/magazine/story/2017/12/16/pentagon-ufo-search-harry-reid-216111.

7. Berlitz, Charles, and William L. Moore. *The Roswell Incident*. Berkley Books, 1997.

8. Blumenthal, Ralph, and Leslie Kean. "No Longer in Shadows, Pentagon's U.F.O. Unit Will Make Some Findings Public." *The New York Times*, July 23, 2020. https://www.nytimes.com/2020/07/23/us/politics/pentagon-ufo-harry-reid-navy.html.

9. Booth, BJ. "The 1965 Kecksburg, Pennsylvania Crash." The Kecksburg UFO Crash, UFO Casebook files. Accessed February 8, 2024. https://www.ufocasebook.com/Kecksburg.html.

10. Booth, BJ. "The Lubbock Lights, 1951." The Lubbock Lights, 1951, UFO Casebook Files. Accessed February 9, 2024. https://www.ufocasebook.com/lubbocklights.html.

11. Brennan, Pat. "Life in the Universe: What Are the Odds? - NASA Science." NASA. Accessed September 10, 2024. https://science.nasa.gov/universe/exoplanets/life-in-the-universe-what-are-the-odds/.

12. Broadcast. *UFO Conspiracy: Hunt for the Truth*. HISTORY, July 7, 2017.

13. C, Kristen. "Message from Outer Space? The Mysterious Indecipherable Script of the Inga Stone." *Ancient Origins Unraveling the Mysterious of the Past*, August 4, 2023. https://www.ancient-origins.net/artifacts-ancient-writings/inga-stone-006442.

14. Cain, Fraser. "How Long Does It Take Sunlight to Reach the Earth?" Phys.org, April 15, 2013. https://phys.org/news/2013-04-sunlight-earth.html.

15. Carey, Thomas J., and Donald R. Schmitt. *Witness to Roswell, 75th Anniversary Edition: Unmasking the Government's Biggest Cover-Up* (Red Wheel Weiser, 2009).

16. Carus, Titus Lucretius, and Rouse W H D. *De Rerum Natura.* (Harvard University Press, 1937).1

7. Cook, Nick. *The Hunt for Zero Point: Inside the Classified World of Antigravity Technology.* (Broadway Books, 2003).

18. Corso, Philip J., and William J. Birnes. *The Day After Roswell.* (Simon & Schuster Inc., 2017).

19. Costa, Telma. "The Language of Inga Stone -a New Theory About the Origen of Phoenician Alphabet-Itacotiara/Brazil." *Oxford University History Society*, March 11, 2020. https://www.academia.edu/42185714/ The_Language_of_Inga_Stone_A_New_Theory_About_the_Origen_of_ Phoenician_Alphabet_Itacotiara_Brazil.

20. Cromie, William. "Alien Abduction Claims Examined." *Harvard Gazette*, January 11, 2024. https://news.harvard.edu/gazette/story/2003/02/ alien-abduction-claims-examined-2/.

21. Curry, Andrew. "Gobekli Tepe: The World's First Temple?" Smithsonian. com, November 1, 2008. https://www.smithsonianmag.com/history/ gobekli-tepe-the-worlds-first-temple-83613665/.

22. "Cydonia Region of Mars." NASA. Accessed January 26, 2024. https:// nssdc.gsfc.nasa.gov/planetary/mgs_cydonia.html.

23. Daniels, Andrew. "Pentagon's UFO Group Is Officially Active, after Years of Secrecy." Popular Mechanics, August 16, 2020. https://www.popularme- chanics.com/military/research/a33614916/pentagon-ufo-task-force-active/.

24. Deighton, Ben. "Belgian Hit UFO Image Was Polystyrene, Says Forger | Reuters." Edited by Allison Williams. Reuters, July 27, 2011. https://www.reuters.com/article/oukoe-uk-belgium-ufo-idAFTRE76Q2DE20110727.

25. Dineen, Hannah. "60 Years Later: The 'alien Abduction' of Betty and Barney Hill ..." News Center Maine, September 19, 2021. https://www.newscentermaine.com/article/features/60-years-later-the-alien-abduction-of-betty-and-barney-hill/97-ae2cf39f-f89c-4ba2-bde0-5cd69c9ae518.

26. Doherty, Peta. "Crop Circle Research Held Back by UFO Conspiracy Links." Scientific Crop Circle Research Held Back by Links to UFO Conspiracies, July 26, 2016. https://www.abc.net.au/news/rural/2016-07-26/scientific-crop-circle-research-held-back-by-ufo-links/7660712.

27. Donovan, Barna William. *Conspiracy films: A tour of dark places in the American conscious.* Jefferson, NC: McFarland, 2011.

28. Editors of Publications International, Ltd. the. "History of the Roswell UFO Incident." HowStuffWorks Science, March 8, 2023. https://science.howstuffworks.com/space/aliens-ufos/history-roswell-incident.htm.

29. "ESA Takes Steps toward Quantum Communications." ESA. Accessed April 3, 2024. https://www.esa.int/Enabling_Support/Preparing_for_the_Future/Discovery_and_Preparation/ESA_takes_steps_toward_quantum_communications.

30. "Ex-Air Force Personnel: Ufos Deactivated Nukes." CBS News, September 28, 2010. https://www.cbsnews.com/news/ex-air-force-personnel-ufos-deactivated-nukes/.

31. "Exoplanet Exploration: Planets beyond Our Solar System," NASA, December 17, 2015, https://exoplanets.nasa.gov/.

32. "Facts about Craniosynostosis." Centers for Disease Control and Prevention, June 28, 2023. https://www.cdc.gov/ncbddd/birthdefects/craniosynostosis.html.

33. "FAQ Lite Webb Telescope/NASA." NASA. Accessed January 26, 2024. https://webb.nasa.gov/content/about/faqs/faqLite.html.

34. "Galilean Telescope." Encyclopædia Britannica. Accessed January 26, 2024. https://www.britannica.com/science/Galilean-telescope.

35. Gershon, Livia. "The 1970s Cow Mutilation Mystery." JSTOR Daily, March 17, 2023. https://daily.jstor.org/the-1970s-cow-mutilation-mystery/.

36. Graves (*uncertainvector), Ryan. "'Ryan Graves UAP Sighting.'" X.Com (Twitter Post). Ryan Graves, August 18, 2023. https://twitter.com/uncertainvector/status/1692586130162475209.

37. Grey, Jeffrey. *Up top: The royal australian navy and Southeast Asian conflicts*, 1955-1972. St. Leonards: Allen & Unwin, 1998.

38. Guerra, John L. *Strange craft: The true story of an Air Force Intelligence Officer's life with ufos*. Tampa, FL: Bayshore Publishing Co., 2018.

39. Guilmartin, John F. "Unmanned Aerial Vehicle." Encyclopædia Britannica, March 7, 2024. https://www.britannica.com/technology/unmanned-aerial-vehicle.

40. Hanson, Robin. "Are We Almost Past It?" The Great Filter, September 15, 1998. https://mason.gmu.edu/~rhanson/greatfilter.html.

41. "Hearing on Unidentified Aerial Phenomena." C-SPAN, July 26, 2023. https://www.c-span.org/video/?529499-1%2Fhearing-unidentified-aerial-phenomena.

42. History.com Editors. "Jimmy Carter Files Report on UFO Sighting | September 18, 1973." History.com. Accessed February 8, 2024. https://www.history.com/this-day-in-history/carter-files-report-on-ufo-sighting.

43. Irving, Rob, and Peter Brookesmith. "Crop Circles: The Art of the Hoax." Smithsonian.com, December 15, 2009. https://www.smithsonianmag.com/arts-culture/crop-circles-the-art-of-the-hoax-2524283/.

44. Kaufman, Marc. "NASA Astrobiology." NASA. Accessed September 10, 2024. https://astrobiology.nasa.gov/about/.

45. Klass, Philip J. "FAA Data Sheds New Light on JAL Pilot's UFO Report." *Skeptical Inquirer*. Issue 11 pp. 322–326. 1987

46. Kuban, Glenn. "The London Hammer:" The London Hammer: An Alleged Out of Place Artifact. Accessed January 28, 2024. http://paleo.cc/paluxy/hammer.htm.

47. "La Catedral Nueva." Catedral de Salamanca, October 1, 2021. https://catedralsalamanca.org/catedral-nueva/#ramos.

48. Lange, Jeva. "30 years later, we still don't know what really happened during the Belgian UFO wave." *The Week*, March 30, 2020. https://theweek.com/articles/905215/30-years-later-still-dont-know-what-really-happened-during-belgian-ufo-wave.

49. Lavery, Kevin. "UFO or Swamp Gas? Mi's 'Close Encounter' 50 Years Later." WKAR Public Media, March 21, 2016. https://www.wkar.org/radio-made-in-michigan/2016-03-21/ufo-or-swamp-gas-mis-close-encounter-50-years-later.

50. Liebermann, Oren. "First on CNN: US Is Receiving Dozens of UFO Reports a Month, Senior Pentagon Official Tells CNN | CNN Politics." CNN, October 18, 2023. https://www.cnn.com/2023/10/18/politics/us-ufo-reports-pentagon/index.html.

51. Lopez, C. Todd. "U.S. Tracking High-Altitude Surveillance Balloon." U.S. Department of Defense, February 2, 2023. https://www.defense.gov/News/News-Stories/Article/Article/3287177/us-tracking-high-altitude-surveillance-balloon/.

52. "Lydia Sleppy Affidavit," Sunrise, September 14, 1993, https://www.sunrisepage.com/roswell/files/witnesses/Sleppy,Lydia.docx.

GENE P. ABEL

53. "Mars 2020 Perseverance Rover." NASA. Accessed January 26, 2024. https://mars.nasa.gov/mars2020/.

54. McMillan, Tim. "Navy UFO | UFO Sightings | the Truth about the Navy's Ufos." Popular Mechanics, November 12, 2019. https://www.popularmechanics.com/military/research/a29771548/navy-ufo-witnesses-tell-truth/.

55. Meares, Hadley. "The Unsolved Mystery of the Lubbock Lights UFO Sightings." History.com, August 24, 2018. https://www.history.com/news/lubbock-lights-ufo-sightings.

56. O'Neill, Ian J. "NASA's Tech Demo Streams First Video From Deep Space via Laser." NASA. Accessed January 28, 2024. https://www.jpl.nasa.gov/news/nasas-tech-demo-streams-first-video-from-deep-space-via-laser.

57. Newsroom, Intel. "Intel Unveils Industry-Leading Glass Substrates to Meet Demand For..." Intel. Accessed February 5, 2024. https://www.intel.com/content/www/us/en/newsroom/news/intel-unveils-industry-leading-glass-substrates.html#gs.4mn4wu.

58. Pester, Patrick. "What did the ancient Egyptian pyramids look like when they were built?" LiveScience, February 5, 2023. https://www.livescience.com/how-egyptian-pyramids-originally-looked.

59. Pilkington, Ed. "We're Not Alone, Says Former NASA Astronaut." *The Guardian*, July 25, 2008. https://www.theguardian.com/science/2008/jul/26/spaceexploration.

60. Potier, Beth. "Starship Memories." *Harvard Gazette*, October 31, 2002. https://news.harvard.edu/gazette/story/2002/10/starship-memories-2/.

61. "Planetary Exploration Timeline." NASA. Accessed January 26, 2024. https://nssdc.gsfc.nasa.gov/planetary/chronology.html.

62. "Project Blue Book (UFO) Part 1 of 1." FBI, December 6, 2010. https://vault.fbi.gov/Project%20Blue%20Book%20%28UFO%29%20/Project%20Blue%20Book%20%28UFO%29%20part%201%20of%201/view.

63. "Project Blue Book - Unidentified Flying Objects." National Archives and Records Administration, September 2020. https://www.archives.gov/research/military/air-force/ufos.

64. Raeburn, Paul. "Scientists Explain Alleged UFO Sighting by Japanese Pilot over Alaska." *AP NEWS*, January 27, 1987. https://web.archive.org/web/20221122233407/https://apnews.com/article/275967ae96c4e21dad2fb5eda04bcb37.

65. Schnabel, Jim. *Round in Circles: Physicists, Poltergeists, Pranksters and the Secret History of the Crop Watchers.* Penguin, 1994.

66. *Secret Access: UFOs on the Record.* Movie, 2011. US: Break Thru Films, n.d.

67. Qiu, Xiaohang, and Chunyu Ding. "Radar Observation of the Lava Tubes on the Moon and Mars." *Remote Sensing* 15, no. 11 (May 30, 2023): 2850. https://doi.org/10.3390/rs15112850.

68. Raeburn, Paul. "Scientists Explain Alleged UFO Sighting by Japanese Pilot over Alaska." *AP NEWS*, January 27, 1987. https://web.archive.org/web/20221122233407/https://apnews.com/article/275967ae96c4e21dad2fb5eda04bcb37.

69. Schneiker, Robert Adam. "The Mystery of Göbekli Tepe: A New Chapter in History." *Skeptic*, May 16, 2022. https://www.skeptic.com/reading_room/gobekli-tepe-mystery-new-chapter-in-history-robert-adam-schneiker/.

70. Segalov, Michael. "Helen Sharman: 'There's No Greater Beauty than Seeing the Earth from up High.'" *The Guardian*, January 5, 2020. https://www.theguardian.com/lifeandstyle/2020/jan/05/astronaut-helen-sharman-this-much-i-know.

71. Sitchin, Zecharia, and Janet Sitchin. *The Anunnaki Chronicles: A Zecharia Sitchin Reader*. Bear & Company, 2015.

72. Sohn, Rebecca. "What Would Happen If You Moved at the Speed of Light?" February 12, 2024. https://www.space.com/what-would-happen-if-you-moved-at-speed-of-light.

73. Smith, Paul Blake. *MO41: The Bombshell before Roswell*. W & B Publishers, 2015.

74. Staff, WIRED. "'UFO Hacker' Tells What He Found." *Wired*, June 21, 2006. https://www.wired.com/2006/06/ufo-hacker-tells-what-he-found/.

75. Steiger, Brad. Project Blue Book: The top secret ufo files that revealed a government cover-up. *Micro Publishing Media, Inc.*, 2019.

76. Stone, Mike. "Pentagon Fails Audit for Sixth Year in a Row" *Reuters*, 2023. https://www.reuters.com/world/us/pentagon-fails-audit-sixth-year-row-2023-11-16/.

77. The Federal Bureau of Investigation, "Majestic 12," FBI Records: The Vault, n.d., https://vault.fbi.gov/Majestic%2012/Majestic%2012%20Part%201%20of%201/view.

78. "The Suspicious Death of James Forrestal," *Gaia*, n.d., https://www.gaia.com/video/the-suspicious-death-of-james-forrestal.

79. Thornton, Jessica E, Kevin M Richards, and J Thomas McClintock. "Raman Spectroscopy and STR Analysis of the Elongated Skulls from the Paracas Mummies of Peru." *Journal of Biotechnology & Bioinformatics Research*, December 31, 2022. https://doi.org/10.47363/jbbr/2022(4)155.

80. Tillman, Nola Taylor, and Jonathan Gordon. "How Big Is the Universe?" January 28, 2022. https://www.space.com/24073-how-big-is-the-universe.html.

81. Tillman, Nola Taylor, and Ailsa Harvey. "What are wormholes?" January 13, 2022. https://www.space.com/20881-wormholes.html#:~:text=Einstein's%20theory%20of%20general%20relativity,affects%20light%20that%20passes%20by.

82. US Department of Defense. "Executive Summary." *AARO*. Accessed September 10, 2024. https://www.aaro.mil/Portals/136/PDFs/AARO_Historical_Record_Report_Vol_1_2024.pdf.

83. US Department of Justice. "London, England Hacker Indicted Under Computer Fraud and Abuse Act For Accessing Military Computers." US Department of Justice, November 12, 2002. https://www.justice.gov/archive/criminal/cybercrime/press-releases/2002/mckinnonIndict.htm.

84. Webster, Senior Airman Erica. "An inside Look at F-35 Pilot Helmet Fittings." Air Force, August 5, 2021. https://www.af.mil/News/Article-Display/Article/2719003/an-inside-look-at-f-35-pilot-helmet-fittings/.

85. Williams, Matt. "What Is the Alcubierre 'Warp' Drive?" Phys.org, January 20, 2017. https://phys.org/news/2017-01-alcubierre-warp.html.

86. Xiao, Zhiyong, Pan Yan, Bo Wu, Chunyu Ding, Yuan Li, Yiren Chang, Rui Xu, et al. "Translucent Glass Globules on the Moon." *Science Bulletin* 67, no. 4 (February 2022): 355–58. https://doi.org/10.1016/j.scib.2021.11.004.

87. Zarenkiewicz, Dan. "Aliens in Alaska." Season 1, Episode 5, March 15, 2021.

End Notes

CHAPTER 1

1   Titus Lucretius Carus and Rouse W H D., De Rerum Natura (Cambridge, MA: Harvard University Press, 1937).

2   "Galilean Telescope," *Encyclopædia Britannica*, accessed January 26, 2024, https://www.britannica.com/science/Galilean-telescope.

3   Zecharia Sitchin and Janet Sitchin, *The Anunnaki Chronicles: A Zecharia Sitchin Reader* (Rochester, VT: Bear & Company, 2015), https://www.amazon.com/Anunnaki-Chronicles-Zecharia-Sitchin-Reader/dp/1591432294.

4   Patrick Pester, "What Did the Ancient Egyptian Pyramids Look like When They Were Built?" LiveScience, February 5, 2023, https://www.livescience.com/how-egyptian-pyramids-originally-looked.

5   Britannica, T. Editors of Encyclopaedia. "How did the Egyptians build the pyramids?." *Encyclopedia Britannica*, October 26, 2018. https://www.britannica.com/question/How-did-the-Egyptians-build-the-pyramids.

6   T.K. Randall, "'Aliens built the pyramids', Tweets Elon Musk,"

Unexplainedmysteries.com, April 4, 2020, 'Aliens built the pyramids', Tweets Elon Musk | Unexplained Mysteries (unexplained-mysteries.com).

7   "Joe Rogan: Are ALIENS statistically possible? Did they build the pyramids?" YouTube Video, January 3, 2023, 7:55, https://www.youtube.com/watch?v=IxbQ-Op9Dr0.

8   "The Great Pyramid: Star Fixed, When?" (PDF). *DIO: The International Journal of Scientific History.* **13** (1): 2–11. December 2003.

9   April Holloway, "Initial DNA Analysis of Paracas Elongated Skull Released – with Incredible Results," Ancient Origins Reconstructing the story of humanity's past, September 28, 2022, https://www.ancient-origins.net/news-evolution-human-origins/initial-dna-analysis-paracas-elongated-skull-released-incredible.

10   April Holloway. "Initial DNA Analysis of Paracas Elongated Skull Released – with Incredible Results." Ancient Origins Reconstructing the story of humanity's past, September 28, 2022. https://www.ancient-origins.net/news-evolution-human-origins/initial-dna-analysis-paracas-elongated-skull-released-incredible.

11   Jessica E Thornton, Kevin M Richards, and J Thomas McClintock, "Raman Spectroscopy and STR Analysis of the Elongated Skulls from the Paracas Mummies of Peru," *Journal of Biotechnology & Bioinformatics Research*, December 31, 2022, 1–8, https://doi.org/10.47363/jbbr/2022(4)155.

12   "Facts about Craniosynostosis," Centers for Disease Control and Prevention, June 28, 2023, https://www.cdc.gov/ncbddd/birthdefects/craniosynostosis.html.

13   Robert Adam Schneiker, "The Mystery of Göbekli Tepe: A New Chapter in History," Skeptic, May 16, 2022, https://www.skeptic.com/reading_room/gobekli-tepe-mystery-new-chapter-in-history-robert-adam-schneiker/.

14   Kristen C, "Message from Outer Space? The Mysterious Indecipherable Script of the Inga Stone," Ancient Origins Reconstructing the story

of humanity's past, August 4, 2023, https://www.ancient-origins.net/artifacts-ancient-writings/inga-stone-006442.

15   Telma Costa, "The Language of Inga Stone - A New Theory about the Origen of Phoenician Alphabet- Itacotiara/Brazil," Oxford University History Society, March 11, 2020, https://www.academia.edu/42185714/The_Language_of_Inga_Stone_A_New_Theory_About_the_Origen_of_Phoenician_Alphabet_Itacotiara_Brazil.

16   Glenn Kuban, "The London Hammer," The London Hammer: An Alleged Out of Place Artifact, accessed January 28, 2024, http://paleo.cc/paluxy/hammer.htm.

17   "La Catedral Nueva," Catedral de Salamanca, October 1, 2021, https://catedralsalamanca.org/catedral-nueva/#ramos.

Chapter 2

18   "Hearing on Unidentified Aerial Phenomena," *C-SPAN*, July 25, 2023, 2 hrs., 18 min., https://www.c-span.org/video/?529499-1/hearing-unidentified-aerial-phenomena.

19   "Hearing on Unidentified Aerial Phenomena," *C-SPAN*, July 26, 2023, https://www.c-span.org/video/?529499-1%2Fhearing-unidentified-aerial-phenomena. Timestamp: 00:13:58

20   "Hearing on Unidentified Aerial Phenomena," *C-SPAN*, July 26, 2023, https://www.c-span.org/video/?529499-1%2Fhearing-unidentified-aerial-phenomena. Timestamp: 00:07:11

21   "Hearing on Unidentified Aerial Phenomena," *C-SPAN*, July 26, 2023, https://www.c-span.org/video/?529499-1%2Fhearing-unidentified-aerial-phenomena. Timestamp: 00:07:42

22   "Hearing on Unidentified Aerial Phenomena," *C-SPAN*, July 26, 2023, https://www.c-span.org/video/?529499-1%2Fhearing-unidentified-aerial-phenomena.

Timestamp: 00:26:04

23   "Hearing on Unidentified Aerial Phenomena," *C-SPAN*, July 26, 2023, https://www.c-span.org/video/?529499-1%2Fhearing-unidentified-aerial-phenomena. Timestamp: 00:27:21

24   "Hearing on Unidentified Aerial Phenomena," *C-SPAN*, July 26, 2023, https://www.c-span.org/video/?529499-1%2Fhearing-unidentified-aerial-phenomena. Timestamp: 01:56:00

25   "Ex-Air Force Personnel: Ufos Deactivated Nukes." *CBS News*, September 28, 2010. https://www.cbsnews.com/news/ex-air-force-personnel-ufos-deactivated-nukes/.

26   The full post and video of of former Navy Lieutenant Ryan Graves's testimony, edited for privacy, can be found online at https://x.com/uncertainvector/status/1692586130162475209?s=20. Accessed September 18, 2024.

Chapter 3

27   Smith, Paul Blake. *MO41: The Bombshell before Roswell*. Kernersville, W & B Publishers, 2015.

28   Ibid.

29   Corso, Colonel Philip, (Retired) with William J. Birnes, *The Day After Roswell* (Gallery Books: 2017).

30   Berlitz, Charles, and William L. Moore, *The Roswell Incident*, Berkley Books, 1997.

31   "Lydia Sleppy Affidavit," Sunrise, September 14, 1993, https://www.sunrisepage.com/roswell/files/witnesses/Sleppy,Lydia.docx.

32   Berlitz, Charles, and William L. Moore, *The Roswell Incident*, Berkley Books, 1997.

33   Spencer, Lawrence R., *Alien Interview* (Lawrence R. Spencer, 2008), pp. 28

34   Twining, Nathan, to George Schulgen, September 23, 1947, Air Materiel Command, AMC Opinion Concerning "Flying Discs, Records of the Army Air Forces, The National Archives.

35   Booth, B.J., "Lieutenant Walter Haut's Deathbed Confession," UFO Casebook, accessed August 21, 2024, https://www.ufocasebook.com/hautconfession.html.

36   Bain, Rachel. "Huge Hydrogen & Oxygen Balloon Explosion." UW-Madison Kaltura MediaSpace. Accessed September 8, 2024. https://mediaspace.wisc.edu/media/Huge+Hydrogen+%26+Oxygen+Balloon+Explosion/1_gjuirv0a.

37   Booth, B.J., "Lieutenant Walter Haut's Deathbed Confession."

38   Area 51 was used for developing and testing aircraft and weapons systems. Its existence was not publicly acknowledged until 2013, when declassified CIA documents referenced the site in connection with spy planes. It has since become synonymous with UFO and alien conspiracy theories. Some believe that the facility houses evidence of extraterrestrial life and technology, possibly linked to the 1947 Roswell incident.

39   The Federal Bureau of Investigation, "Majestic 12," FBI Records: The Vault, n.d., https://vault.fbi.gov/Majestic%2012/Majestic%2012%20Part%201%20of%201/view.

40   Donovan, Barna William, *Conspiracy Films: A Tour of Dark Places in the American Conscious* (Jefferson, NC: McFarland, 2011).

41   Dolan, Richard, "The Suspicious Death of James Forrestal," Aerial Phenomena with Richard Dolan, Gaia, Season 1, Episode 9, accessed August 21, 2024, https://www.gaia.com/video/the-suspicious-death-of-james-forrestal.

42   "A Little Help with Norma Gardner," http://kevinrandle.blogspot.com/2016/02/a-little-help-with-norma-gardner.html, February 22, 2016, http://kevinrandle.blogspot.com/2016/02/a-little-help-with-norma-gardner.html.

CHAPTER 4

43   Liebermann, Oren, "First on CNN: US Is Receiving Dozens of UFO Reports a Month, Senior Pentagon Official Tells CNN | CNN Politics," CNN, October 18, 2023, https://www.cnn.com/2023/10/18/politics/us-ufo-reports-pentagon/index.html.

44   Lopez, C. Todd, "U.S. Tracking High-Altitude Surveillance Balloon," U.S. Department of Defense, February 2, 2023, https://www.defense.gov/News/News-Stories/Article/Article/3287177/us-tracking-high-altitude-surveillance-balloon/.

45   "Project Blue Book - Unidentified Flying Objects," National Archives and Records Administration, September 2020, https://www.archives.gov/research/military/air-force/ufos.

46   Steiger, Brad, Project Blue Book: The Top Secret Ufo Files That Revealed a Government Cover-Up (Micro Publishing Media, Inc., 2019).

47   Meares, Hadley, "The Unsolved Mystery of the Lubbock Lights UFO Sightings," HISTORY, January 10, 2020, https://www.history.com/news/lubbock-lights-ufo-sightings.

48   Booth, B.J., "The Lubbock Lights, 1951," The Lubbock Lights, 1951, UFO Casebook Files, accessed February 9, 2024, https://www.ufocasebook.com/lubbocklights.html.

49   Booth, B.J., "The 1965 Kecksburg, Pennsylvania Crash," The Kecksburg UFO Crash, UFO Casebook files, accessed February 8, 2024, https://www.ufocasebook.com/Kecksburg.html.

50   Lavery, Kevin, "UFO or Swamp Gas? Mi's 'Close Encounter' 50 Years Later," WKAR Public Media, March 21, 2016, https://www.wkar.org/radio-made-in-michigan/2016-03-21/ufo-or-swamp-gas-mis-close-encounter-50-years-later.

51   UFO Conspiracy: Hunt for the Truth (HISTORY, July 7, 2017).

52   History.com Editors, "Jimmy Carter Files Report on UFO Sighting | September 18, 1973," History.com, accessed February 8, 2024, https://www.history.com/this-day-in-history/carter-files-report-on-ufo-sighting.

53   Guerra, John L., *Strange Craft: The True Story of an Air Force Intelligence Officer's Life with UFOs* (Tampa, FL: Bayshore Publishing Co., 2018).

54   Raeburn, Paul, "Scientists Explain Alleged UFO Sighting by Japanese Pilot over Alaska," AP NEWS, January 27, 1987, https://web.archive.org/web/20221122233407/https://apnews.com/article/275967ae96c4e21dad2fb5eda04bcb37.

55   Klass, Philip J. "FAA Data Sheds New Light on JAL Pilot's UFO Report." *Skeptical Inquirer*. Issue 11 pp. 322–326. 1987

56   *Secret Access: UFOs on the Record*, DVD (US: Break Thru Films, n.d.).

57   Deighton, Ben, "Belgian Hit UFO Image Was Polystyrene, Says Forger | Reuters," ed. Allison Williams, Reuters, July 27, 2011, https://www.reuters.com/article/oukoe-uk-belgium-ufo-idAFTRE76Q2DE20110727.

58   Lange, Jeva, "30 Years Later, We Still Don't Know What Really Happened during the Belgian UFO Wave," theweek, March 30, 2020, https://theweek.com/articles/905215/30-years-later-still-dont-know-what-really-happened-during-belgian-ufo-wave.

59   WIRED Staff, "'UFO Hacker' Tells What He Found," Wired, June 21, 2006, https://www.wired.com/2006/06/ufo-hacker-tells-what-he-found/.

60   US Department of Justice, "London, England Hacker Indicted Under Computer Fraud and Abuse Act For Accessing Military Computers," US Department of Justice, November 12, 2002, https://www.justice.gov/archive/criminal/cybercrime/press-releases/2002/mckinnonIndict.htm.

61   BBC Staff, "Hacker Gary McKinnon Turns into a Search Expert," BBC News, July 28, 2014, https://www.bbc.com/news/technology-28524909.

62   McMillan, Tim, "Navy UFO | UFO Sightings | the Truth about the Navy's Ufos," Popular Mechanics, November 12, 2019, https://www.popular-mechanics.com/military/research/a29771548/navy-ufo-witnesses-tell-truth/.

63   United States Department of Defense. "Executive Summary." AARO. Accessed September 10, 2024. https://www.aaro.mil/Portals/136/PDFs/AARO_Historical_Record_Report_Vol_1_2024.pdf.

64   Bender, Bryan, "The Pentagon's Secret Search for UFOs - Politico Magazine," Politico, December 16, 2017, https://www.politico.com/magazine/story/2017/12/16/pentagon-ufo-search-harry-reid-216111.

65   Daniels, Andrew, "Pentagon's UFO Group Is Officially Active, after Years of Secrecy," Popular Mechanics, August 16, 2020, https://www.popularmechanics.com/military/research/a33614916/pentagon-ufo-task-force-active/.

66   "A Little Help with Norma Gardner," http://kevinrandle.blogspot.com/2016/02/a-little-help-with-norma-gardner.html, February 22, 2016

67   Blumenthal, Ralph, and Leslie Kean, "No Longer in Shadows, Pentagon's U.F.O. Unit Will Make Some Findings Public," The New York Times, July 23, 2020, https://www.nytimes.com/2020/07/23/us/politics/pentagon-ufo-harry-reid-navy.html.

CHAPTER 5

68   Dineen, Hannah, "60 Years Later: The 'alien Abduction' of Betty and Barney Hill ...," News Center Maine, September 19, 2021, https://www.newscentermaine.com/article/features/60-years-later-the-alien-abduction-of-betty-and-barney-hill/97-ae2cf39f-f89c-4ba2-bde0-5cd69c9ae518.

69   Cromie, William, "Alien Abduction Claims Examined," Harvard Gazette, January 11, 2024, https://news.harvard.edu/gazette/story/2003/02/alien-abduction-claims-examined-2/.

70   Potier, Beth, "Starship Memories," Harvard Gazette, October 31, 2002, https://news.harvard.edu/gazette/story/2002/10/starship-memories-2/.

71   Zarenkiewicz, Dan, "Aliens in Alaska," episode, March 15, 2021.

72   Gershon, Livia, "The 1970s Cow Mutilation Mystery," JSTOR Daily, March 17, 2023, https://daily.jstor.org/the-1970s-cow-mutilation-mystery/.

73   FBI, "Animal Mutilation," FBI, December 6, 2010, https://vault.fbi.gov/Animal%20Mutilation.

74   Irving, Rob, and Peter Brookesmith, "Crop Circles: The Art of the Hoax," Smithsonian.com, December 15, 2009, https://www.smithsonianmag.com/arts-culture/crop-circles-the-art-of-the-hoax-2524283/.

75   BBC Editors, "New Data Shows Wiltshire Has Most Crop Circles in England," BBC News, July 22, 2023, https://www.bbc.com/news/uk-england-wiltshire-66245271.

76   Schnabel, Jim, *Round in Circles: Physicists, Poltergeists, Pranksters and The Secret History of the Cropwatchers* (Penguin, 1994).

77   Doherty, Peta, "Crop Circle Research Held Back by UFO Conspiracy Links," Scientific Crop Circle Research Held Back by Links to UFO Conspiracies, July 26, 2016, https://www.abc.net.au/news/rural/2016-07-26/scientific-crop-circle-research-held-back-by-ufo-links/7660712.

78   Carey, Thomas J., and Donald R. Schmitt, *Witness to Roswell, Revised and Expanded Edition: Unmasking the Government's Biggest Cover-Up* (Red Wheel Weiser, 2009).

79   Ltd. the Editors of Publications International, "History of the Roswell UFO Incident," HowStuffWorks Science, March 8, 2023, https://science.howstuffworks.com/space/aliens-ufos/history-roswell-incident.htm.

80   Corso, Philip, *The Day after Roswell* (Simon and Schuster, 2012). 47-50.

81   Webster, Senior Airman Erica, "An inside Look at F-35 Pilot Helmet Fittings," Air Force, August 5, 2021, https://www.af.mil/News/Article-Display/Article/2719003/an-inside-look-at-f-35-pilot-helmet-fittings/.

82   Cook, Nick, *The Hunt for Zero Point: Inside the Classified World of Antigravity Technology* (Broadway Books, 2003).

83   Intel Newsroom, "Intel Unveils Industry-Leading Glass Substrates to Meet Demand For...," Intel, accessed February 5, 2024, https://www.intel.com/content/www/us/en/newsroom/news/intel-unveils-industry-leading-glass-substrates.html#gs.4mn4wu.

84   "NASA's Tech Demo Streams First Video from Deep Space via Laser," NASA, accessed January 28, 2024, https://www.jpl.nasa.gov/news/nasas-tech-demo-streams-first-video-from-deep-space-via-laser.

CHAPTER 6

85   "Cydonia Region of Mars," NASA, accessed January 26, 2024, https://nssdc.gsfc.nasa.gov/planetary/mgs_cydonia.html.

86   "Planetary Exploration Timeline," NASA, accessed January 26, 2024, https://nssdc.gsfc.nasa.gov/planetary/chronology.html.

87   Zhiyong Xiao et al., "Translucent Glass Globules on the Moon," *Science Bulletin* 67, no. 4 (February 2022): 355–58, https://doi.org/10.1016/j.scib.2021.11.004. https://www.sciencedirect.com/science/article/abs/pii/S2095927321006964 (source needs rights purchased)

88   Qiu, Xiaohang, and Chunyu Ding. 2023. "Radar Observation of the Lava Tubes on the Moon and Mars" *Remote Sensing* 15, no. 11: 2850. https://doi.org/10.3390/rs15112850 https://www.mdpi.com/2072-4292/15/11/2850.

89   "Mars 2020 Perseverance Rover," NASA, accessed January 26, 2024, https://mars.nasa.gov/mars2020/.

90   "FAQ Lite Webb Telescope/NASA," NASA, accessed January 26, 2024, https://webb.nasa.gov/content/about/faqs/faqLite.html.

91   "About the Roman Space Telescope," NASA, n.d., https://www.nasa.gov/content/goddard/about-nancy-grace-roman-space-telescope.

92   Kaufman, Marc. "NASA Astrobiology." NASA. Accessed September 10, 2024. https://astrobiology.nasa.gov/about/.

93   Carey, Thomas J., and Donald R. Schmitt, *Witness to Roswell, Revised and Expanded Edition: Unmasking the Government's Biggest Cover-Up* (Red Wheel Weiser, 2009).

94   Tillman, Nola Taylor, and Jonathan Gordon, "How Big Is the Universe?" Space.com, January 28, 2022, https://www.space.com/24073-how-big-is-the-universe.html.

95   "Exoplanet Exploration: Planets beyond Our Solar System," NASA, December 17, 2015, https://exoplanets.nasa.gov/.

96   Brennan, Pat. "Life in the Universe: What Are the Odds? - NASA Science." NASA. Accessed September 10, 2024. https://science.nasa.gov/universe/exoplanets/life-in-the-universe-what-are-the-odds/.

97   Aldrin, Buzz, "Buzz Aldrin on the Existence of Extraterrestrials in Our Galaxy | Forbes," YouTube, October 23, 2021, https://www.youtube.com/watch?v=lNJpVDdUeUU.

98   Pilkington, Ed, "We're Not Alone, Says Former NASA Astronaut," *The Guardian*, July 25, 2008, https://www.theguardian.com/science/2008/jul/26/spaceexploration.

99   Segalov, Michael. "Helen Sharman: 'There's No Greater Beauty than Seeing the Earth from up High.'" *The Guardian*, January 5, 2020. https://www.theguardian.com/lifeandstyle/2020/jan/05/astronaut-helen-sharman-this-much-i-know.

100   Sohn, Rebecca, "What Would Happen If You Moved at the Speed of Light?" Space.com, February 12, 2024, https://www.space.com/what-would-happen-if-you-moved-at-speed-of-light.

101   Williams, Matt, "What Is the Alcubierre 'Warp' Drive?" Phys.org, January 20, 2017, https://phys.org/news/2017-01-alcubierre-warp.html.

102   Tillman, Nola Taylor, and Ailsa Harvey, "What Are Wormholes?" Space.com, January 13, 2022, https://www.space.com/20881-wormholes.html#:~:text=Einstein's%20theory%20of%20general%20relativity,affects%20light%20that%20passes%20by.

103   "ESA Takes Steps toward Quantum Communications," ESA, accessed April 3, 2024, https://www.esa.int/Enabling_Support/Preparing_for_the_Future/Discovery_and_Preparation/ESA_takes_steps_toward_quantum_communications

104   Hanson, Robin. "Are We Almost Past It?"Accessed September 10, 2024. https://mason.gmu.edu/~rhanson/greatfilter.html.

www.ingramcontent.com/pod-product-compliance
Lightning Source LLC
Chambersburg PA
CBHW050842270326
41930CB00019B/3443